# A SAGA OF EVOLUTION AND LEGENDS OF ENVIRONMENTAL DISASTERS IN THE HISTORY OF MANKIND

Dr. NIKHIL CHANDRA MISRA

INDIA • SINGAPORE • MALAYSIA

**Notion Press**

No.8, 3rd Cross Street
CIT Colony, Mylapore
Chennai, Tamil Nadu – 600004

First Published by Notion Press 2021
Copyright © Dr. Nikhil Chandra Misra 2021
All Rights Reserved.

ISBN 978-1-63633-512-4

This book has been published with all efforts taken to make the material error-free after the consent of the author. However, the author and the publisher do not assume and hereby disclaim any liability to any party for any loss, damage, or disruption caused by errors or omissions, whether such errors or omissions result from negligence, accident, or any other cause.

While every effort has been made to avoid any mistake or omission, this publication is being sold on the condition and understanding that neither the author nor the publishers or printers would be liable in any manner to any person by reason of any mistake or omission in this publication or for any action taken or omitted to be taken or advice rendered or accepted on the basis of this work. For any defect in printing or binding the publishers will be liable only to replace the defective copy by another copy of this work then available.

# Contents

*Foreword* ..........................................................................5

*Preface* .............................................................................7

*Acknowledgements* ........................................................13

**Chapter 1**   An Account of Universe, Solar System, Moon and Planet Earth ............................................19

**Chapter 2**   Mother Earth and the Antiquity of Life on the Planet ............................................................31

**Chapter 3**   A Version of the Birth and Evolution of Human Race on Planet Earth ..................................51

**Chapter 4**   Environment, Its Elementary Concepts and the Veritable Concepts of Free Commodities, *Carrying Capacity* and Sustainable Development ..................................93

**Chapter 5**   Empathy of the Climate Change, Global Warming and Environmental Damage .......109

**Chapter 6**   Dawn of Civilisation: Indus Valley Civilisation, the Master Traders in Erstwhile Indian Sub-Continent .............................................121

| **Chapter 7** | Dawn of Civilisation: Sumerian in the 'Cradle of Human Civilisation', Did Not Disappear Suddenly ................................................. 143 |
| --- | --- |
| **Chapter 8** | Calamitous Occurrences with the Mysterious Mayans in Outer World ....................... 155 |
| **Chapter 9** | The Appalling Events in Rapa Nui, an Exemplary Case of Gradual Despair and Sustainability ................................................... 169 |

*Bibliography of References & Further Readings* .......................... 197

*Index of Words* ................................................................................ 215

*About the Author* ............................................................................ 237

# Foreword

### THE PERSPECTIVE… AN APERTURE

Whether it was the deep ocean or the soaring skies where they took birth, whether it was damp, dark earth or the rugged, sun-beaten rock where they took roots, every child of Mother Nature possesses a deeply-ingrained, almost symbiotic relationship with the rest of its siblings—a relationship thriving on the perfect harmony that prevails throughout the universe known to us.

However, this balance is feeble and the slightest shift might put the whole system through an utter and complete breakdown.

Let me present a small fable to drive home my point.

An old fish lived in the ocean where she was born and where she grew up and where she was content to spend the rest of her life. One day, however, the water that had been the breath of life for her started to taste like poison. She choked and suffocated. She saw the rest of her kin die and rot away. Befuddled, she went to the tree that stood majestically on the shore for counsel.

'I know what bothers you, Fish. It is Man—the hairless ape who is behind this disorder. He cut my brother down. It is he who made the earth shake and the ocean boil.'

A bird sitting on one of the tree's branches joined in, 'He demolished our home. He took away our food. He put great metal birds in the skies that kill us with their terrible talons.'

'Go to him, O wise Fish. Speak with him of his terrible deeds. Maybe he will relent. Maybe we can still be saved.'

The old fish had hope but fear eclipsed it soon. She was aware of man's abilities and his intelligence. And she knew that despite his conscientious heart, his brain could never let go of the greed that had become the foundation of his very nature. She was reminded of the terrible fate that befell the Indus Valley people, the Mayans and the Sumerians long ago or the terrifying end of the people of Rapa Nui who are said to have resorted to eating their kin to survive.

Disappointed, the old fish fondly recalled the illustrious Lady Burton and begged pardon for reciting her popular, guiltless maxim albeit with a tiny modification. 'He who knows and knows that he knows; he is wise—follow him.' She said 'Human is not'.

— **Manas Misra**

# Preface

The science of nature has been in existence since the universe started to undertake its incredible journey of becoming what it is today, through several formidable events. This science, comprising several disciplines, has enlightened us, humans, about the much-needed necessity to identify the mistakes we have been committing in the recent past. And when the wrong deeds are identified, we the humans have become more intelligent and aware of what is exactly needed as etiquettes—the environmental etiquettes. We also got to know the intricacies. It is said that the glorified legacy that is nurtured adorably leads to the bright future via the opulent present. By now we have acquired an incredible ability to learn about our environment and to communicate what we learn to mankind across the planet.

Evolution including the biological evolution is a perpetual phenomenon. It laid the foundation for a discipline that provides a platform for continuous research. Most of the predictions which remained prophecies for quite a period were later on confirmed by observations and experiments. The birth of the universe is viewed by different schools of scientists and thinkers in a somewhat diverse manner, imposing cautions. All have one thing in common and that is its tiny size at the proverbial beginning. The path-breaking work on the study of Nebula and galaxy by Edwin Hubble not only delivered the much-needed path but also provided the necessary basis for further work. The *Big Bang Theory* was born. The universe expanded to the size of the Milky Way. Our story proceeds, the earth, moon and sun would spin off from the rich debris left over from one of the supernovae. Debris grew to critical mass, burnt, and the sun was born. The leftover debris created planets including the earth. The atmosphere was formed after about 500 million years. Aryabhatt, the Indian wizard, inventor of "zero", made a monumental contribution in the field. Vedic literature of ancient India was probably the first to propose the universe as 'heliocentric'. Polish mathematician and astronomer, Nicholas Copernicus propounded 'heliocentric' hypothesis. More contributions came along in later years. The American astronauts launched and repaired in space itself, the Hubble telescope. The moon was created by the glancing collision of Theia, a Mars-size planet with earth about 4.5 billion years ago. The fiercely-debated Giant Impact Hypothesis was propounded.

Extinction and rebuilding occurred several times. After the essentials developed, life came up in the form of micro-sized unicellular bacteria in the midst of many such occurrences. Water and Minerals were presumably brought along by meteoroids and asteroids with them. Life and essentials did not ascend all at once. Biological evolution followed, a divergent array of species was formed, probably from one or a few common progenitors. When did humans originate and from whom? When did they appear to occupy the planet earth? What journey did the process of human evolution undertake to reach the Homo group? Legendary Leaky family and amazing people with an outstanding knack for research like Dr. Jane Goodall made path-breaking contributions in the fields of anthropology and biology. African continent became the cradle of humanity.

In due course, the human society learnt the finesse of the environment, its degradation, global warming and climate change.

The narrative shifts to become more historical reflecting that the damage caused to the environment can completely wipe out advanced civilisations also. With a penchant for turning the leaves of antiquity, the instances are aplenty in the history of mankind.

The valleys of River Indus and revered River Saraswati, as well as some other places, were inhabited by a civilisation from about 3300 BCE, now anticipated to be much earlier than 5000 BCE or even earlier than 8000 BCE, flourishing till about 1300 BCE before either migrating to different places in groups and/or intermingling with others. The master traders, inventors of weights and measures, the traders of gemstones and exceptional urban planners, the Indus Valley people exterminated unceremoniously probably due to the vagaries of drought and similar derogatory manifestations resulting from what we now call climate change.

The basin area that lay between the Tigris and the Euphrates rivers in southern Mesopotamia, later Babylonia and finally southern Iraq acquired the epithet of 'Cradle of civilisation', witnessed the inception of Sumerians. Credited for 39 'firsts', they built big cities, practised medicine, initiated funeral chants, started schools, developed a highly advanced system of irrigation for farming, built big cities and virtually initiated the moral ethics. Surprisingly, such an advanced pioneering civilisation perished due to the excess of water that spoiled their cultivation strategies gradually exterminating, facing the shortage of food supplies.

The region of *Mesoamerica* comprising Nicaragua, Costa Rica, El Salvador, Guatemala, Belize and central-to-southern México saw the inception of many civilisations but Mayans stood conspicuously in the mob. The Maya are a congregation of an indigenous group of people living in the Central Americas. They speak many languages. They built 40 great cities. They adopted agriculture and started the domestication of maize, beans, spices like vanilla and fruits like Avocado. Maya practised a refined system of governance and the clans were ruled by a respected king whom they called 'Kuhul Ajaw'. The Maya people worshipped various gods, directed to nature. They knew the value of sun, called Kuhul Ajaw. They developed a complex calendar system. They extensively used *cenotes* for fresh water for various purposes and drinking. The Maya Civilisation adopted the practice of human sacrifice, for religious purposes. In their case also the collapse probably happened due to shortage of water, a prevalent eccentricity in environmental command, the rife broke out within and among the clans and cities, taking the toll of an advanced civilisation.

An island, covert in locus standee, shaped almost like a triangle, known by the name of Easter Island also Rapa Nui or Paaseiland is a piece of elevated land in midst of Pacific Ocean a few thousand kilometres away from any land. This island, which is part of the Polynesian triangle was originally settled by Polynesians sailors between 700—800 CE. A club of three volcanoes, the island was lush green with palm groves all over the land. The soil on Easter Island is the source of *Rapamycin* used to prevent the rejection of organ transplant and has an anti-ageing effect also. The islanders had a strong belief in mythical fables and were passionate about carving large human sculptures. Carving of such large rock sculptures and transport to various places across the 163.6-square-kilometres-island are unexplained even today. The islanders cultivated sweet potatoes, hunted *Dolphins* in the sea and thrived gracefully with nature. The *Dolphins* were hunted in the open sea by harpoons, using large canoes. Unscrupulously, they cut the palm trees to build *canoes* and also to transport the Mo'ai (Moai), as they called their sculptures, to various destinations on the island. They were virtually not left with any agriculture as all the palms had been cut and the mineral-rich soil and humus had eroded. The fine balance of sustainability was profoundly damaged. The Polynesian rat was another menace for palm nuts and saplings which also halted the next line of plantation growth. Jared Diamond propounded his famous *'ecocide'* theory. It was a copybook model for disruption in the fine balance of sustainability that Easter Island presented.

Dr. Jason Ur of Harvard University points out the inability of the human populace on the planet and is rather unfair when he says, "When we excavate the remains of past civilisations, we rarely find any evidence that they made any attempts to adapt in the face of a changing climate. I view this inflexibility as the real reason for the collapse." Let's arbiter it carefully. One can adapt and survive. Adapting to the changes is probably an instinct of living beings. What when certain components of the environment actually do not have substitutes? Oxygen for breathing, for instance. As scientists continue to turn up even more signs of collapsed civilisations, they are finding plenty of evidence that climate shifts are to blame to a considerable extent for the decline in many cases. Those links offer the opportunity to protect the future of our own society by learning from the mistakes of our ancestors.

The chronicle is before you which might engender a child-like fascination as you delve deeper and further into it.

# Acknowledgements

*At the outset, I bow to the celestial power that we name in a large variety of ways but mean one.*

It gives me immense pleasure to express my deep gratitude to my respected, adorable forefathers who inadvertently and otherwise encouraged me to undertake challenges in life related to learning—be it complexities of life, conducting matters with full dedication and tenacity, learning of languages, science, humanities and cart unconditional admiration and reverence for legacy. Ambling through the journey of learning during working on the book has been a very pleasant experience, being sandwiched within a closely-knit family of my sons—Manas and Vaibhav, daughter-in-law Swati, my wife Abha and added afresh, loving, tenacious, truly a ball of energy, my grand-daughter beloved Aarna; all encouraged me to keep going in their respective way. What I have learnt and incorporated in the 'A SAGA…' is just anecdotal, having attained with great pleasure. The help that I got from these high-tech computer connoisseurs and truly dedicated experts in their arena of profession has been commendable. Particularly, more indulgent, Manas the younger one, loaded with literary excellence, without whose timely rescue, probably the book would have been left unfinished. The contribution of Swati-Vaibhav and Manas in terms of thought process, wife Abha for making life more comfortable; have undeniably, conjointly helped me in creating what is now in your hands. The unstoppable entertainment provided by loveable Aarna made the intricacies of the subject easier and the spirit soaring high.

Some figures, graphics and photographs are either obtained directly or indirectly from various sources including the internet, which have indeed been inordinate support for advancing my own ideas. I place on records my sincere gratitude to the publishers of such material, the related literature and the authors and authoresses from which they have been obtained without any intention of overtly infringing the copyrights and/or distorting the monumental work those people have carried out.

As regards the book, the core theme, though scientific and historical, attempts to drive home a message. The civilisations flourished to reach the pinnacle, imbibing the finery in life, taught many things to the human society, but true pioneers only got exterminated. The mankind does not seem to worry about what could happen, if the human society does not change its practices and attitude towards the environment, ecology, climate change; and does not strike a sustainable balance among various dominions.

I would place on record my deep gratitude to reverend Shri Yadu Pati Ji Singhania, the principal in a premier business, industrial conglomerate, JK Organisation, a renowned technocrat, a naturalist, a leader of class, a philanthropist, a loving guardian, a highly respected industrialist, and my employer during my tenure with the organisation that he adorably nurtured. His untimely, unceremonious departure from this world recently, leaves a dent that is beyond repair.

I avail myself of the opportunity to express my sincere gratitude to the publisher and their immensely cooperative, able and vibrant team that helped me in providing the perfect shape to the book it is in now.

Chances are you have inadvertently got the 'A SAGA…' in your hand. Chances are you may have come across a favourable review on it that invoked your curiosity. Chances are now that you may like to read it also. Chances are you would like to endeavour deeper and higher. Chances are you may place me to sit in the dock of a courtroom. Chances are you may place me to stand in the witness box. Chances are your choice of place tallies mine.

Chances are you like to delve deeper and higher in the contents and subject matter of the 'A SAGA…' Chances are that I am in luck if you do something like that.

Gratified reading…….

Gratefully.

*A Student, who jumped the Graduation and obtained a PhD from Cambridge, a true advocate for the environmental regime on the planet.*

"Fortunately, nature is amazingly resilient: places we have destroyed, given time and help, can once again support life, and endangered species can be given a second chance. And there is a growing number of people, especially young people, who are aware of these problems and are fighting for the survival of our only home, Planet Earth. We must all join that fight before it is too late."

**– Dr. Jane Goodall**

It is up to us to save the world for tomorrow. It is up to you and me.

**– Dr. Jane Goodall**

"If we all get together, we can truly make a difference, but we must act now. The window of time is closing."

**– Dr. Jane Goodall**

*(Dame Jane Morris Goodall, the English primatologist, anthropologist, world's foremost expert on chimpanzees, member of Nonhuman Rights Project, UN Messenger of Peace, an honorary member of World Future Council, one among 100 most influential people in the world in 2019, named by Time, founder of Jane Goodall Institute and Roots & Shoots and many more… A legend.)*

"Stunningly beautiful nature, the 'Nisarg' (निसर्ग), the 'Nisargtah' (निसर्गतः) the 'Nisargah' (निसर्गः), the 'Prakriti' (प्रकृति) the 'Kudrat', the "la nature", the 'naturae', lives on with sombre grumbles." Save it...

Dr. Nikhil Chandra Misra, *"A Saga of Evolution and Legends of Environmental Disasters in the History of Mankind"*

Warm-Heartedly saluting the sagacity, tenacity, determination, commitment, compassion for nature—saluting Dr. Jane Goodall...

\* \* \* \* \*

# Chapter 1

# An Account of Universe, Solar System, Moon and Planet Earth

Riddles form. They form to be solved—sometimes demonstrated and applauded, sometimes hypothesised and alluded and sometimes just doomed. Probably it is inevitable—but certainly not when it is intricate. It is utterly plausible that earth's simplest life forms, which are little more than tiny bags of elements, the chemicals, formed from the nuclear waste of the stars acting as natural reactors operative in the cosmos, could have remained unchanged. Our complex cells, with their internal compartmental structures and complex support mechanism, their transportation fleets, their intricate machinery, might never have arisen. But then one fine day, about a few billion years ago, a fluke manifestation occurred, then another and yet another. Organic molecules formed from the reactions of the most ubiquitous of materials—rock, water and carbon-di-oxide—and they are thermodynamically close to inevitable. The result? You and I. The existence traverses through the convoluted web of a countless number of factors. Travelling back in time is enormously exciting. It is the times ahead that life does not know about. This is exciting too and probably fearsome to a great extent. Almost mid-way it is a grave road bump—the domain of environment and climate change, leading to the disruption of nature—all fundamentally self-destructing. Stunningly beautiful nature, the *"Nisarg"* (निसर्ग), the *'Nisargtah'* (निसर्गतः) the *'Nisargah'* (निसर्गः), the *'Prakriti'* (प्रकृति) the

"*Kudrat*", the "*la nature*", the "*natura*", lives on with sombre grumbles. Let's delve into the primordial, come along—it is through the science and secrets of luck, randomness and probability that we conjointly undertake this exhilarating journey.

We are an integral part of the cosmos, vital in constructing the shape it is in today and cannot be dissociated from it. The substance related to the astronomical concepts forms the subject matter of astronomy, a complex science. All that is comprehended has taken its origin from this concept. Broadly it studies the universe, solar system, planets, galaxies, meteoroids, asteroids and several other astronomical bodies, often thought to be of the status of celestial origin. The universe is the largest of these three astronomical concepts, the other two being the solar system and the galaxy. All these things are included within the realm of the universe and everything known to man is contained within it. When there is talk of the origin and evolution of the universe, we are talking of our own origin and evolution. The most accepted hypothesis tells us that the entire universe of which solar system, the stars, the earth and moon are a part, came into existence in the form of a marble or even smaller. In the beginning, there was nothing, no time or any such thing. It began with a tiny speck of light. It was infinitely hot with all the space inside this tiny ball of fire. Everything—all the matter, all the energy, that we see today were within that ball of the size of a marble, even smaller than it, probably a single atom. The American Astronomer, Edwin Hubble, also nicknamed "pioneer of distant stars" by his brilliant work and analysis changed the very perception about our universe, helping to lay the foundation of the popular *Big Bang Theory*. The thought that the universe was expanding proved to be the game-changer which meant that it must have been small at some point of time in the past. This indisputably came up as a result of Hubble's study of Androbinan Nebula and a galaxy—other than the one we probably thought we knew about. The famous theory of relativity propounded by Albert Einstein earlier provided the footing.

American Theoretical Astrophysicist David Nathaniel Spergel, an expert in *Big Bang Theory*, opined that this theory does not explain how the universe began but explains how it evolved. Immediately, as immediate as a human mind can conceive, probably in less than a trillion trillionth of a second, the super-hot infant universe which was very unstable then, underwent an enormous growth spurt and started expanding and with it, space was also expanding, faster than the speed of light. After a trillion trillionth of a second, the universe was just big enough to fit in our palm; in the next fraction of a second it matched the size of Mars and the following fraction of a second saw the baby universe acquire the size of about 90 times the earth. Till then, the universe did not contain matter. It was just a ball of energy—enormous energy. Einstein's famous equation $E=MC^2$ explains that matter and energy are interchangeable. However, the universe not only produced matter from energy but also created antimatter. They were arch-rivals, by obliterating each other they would have annulled each other and probably no matter would have been formed had the situation been so. As a consequence, the universe would have remained full of energy and the galaxies, stars or even we would never have come to exist. Fortunately, the matter exceeded the antimatter, i.e. for every one million particles of antimatter formed, one million and one particles of matter were formed—thus matter came into existence.

Scientists created the atmosphere that was exactly akin to what it was about one-millionth of a second after the Big Bang in the Relativistic Heavy Ion Collider (RHIC) in Brookhaven National Laboratory located in Long Island, USA. Their experiments changed the basic thought. The particles on collision did not form a gas but a super-hot liquid with no viscosity at all. The temperature was a 100 million times hotter than the sun. This liquid existed for a tiny fraction of a second. Nevertheless, the universe was in turmoil. The subatomic particles were racing incredibly fast and colliding with each other in the young universe. Because of the mind-boggling speed the particles collided with each other, the electrons were not able to bond to form the atoms. During the next about 300,000 years or more such a situation continued. After a lapse of about 380,000 years following the Big Bang, the universe had become the size of the Milky Ways and its temperature had also come down from a billion degree Fahrenheit to just a few thousand. Later Dr. Arno Penzias of Bell Laboratory conducted research to give an undeniable account of the birth of light which became one of the most important scientific discoveries on earth and helped in solving the problem related to understanding the development of the universe. As time flew, the gases condensed to form stars. Stars constituted the galaxies—like the Milky Ways. Galaxy is a system comprising of solar systems and stars. They are held together by gravity. In galaxies, vast empty spaces are present separating solar systems. Our earth and solar system together with presumably 200 billion stars form a galaxy, called the Milky Way. Solar systems orbit around the stars that form these galaxies. Within the galaxies, the spaces with intense gravitational pulls from which not even light can escape are called black holes. Such sections of space are also present elsewhere in the space.

British astronomers, Fred Hoyle, one of the greatest scientists of the twentieth century refuted that the universe was formed as a result of a single explosion. He was candid when he said, "It is no more likely that our world has evolved out of chaos than that a hurricane, blowing through a junkyard, should create a Boeing". He wondered where the elements heavier than hydrogen and helium came from. The idea put forth was that inside the stars something was going on. The stars acted like nuclear reactors, something almost akin to an atom bomb exploding in slow motion but many billions of times more powerful and their nuclear waste created new elements. This became more coherently accepted by the scientists later on when studies were conducted on the light emitted by stars. On heating, each element emits light at a particular frequency. Let's take sunlight for example. When broken up into spectrum it breaks like a bar code, each colour corresponding to different elements, each of them has a specific colour. Based on this, different elements are identified. For example, hydrogen mainly emits red colour. The Hubble Space Telescope was launched by NASA in the year 1989 to discover the mysteries of an earlier universe. Very appropriately, they called it a window into space. The focus was to study the older stars if they are still making new elements. This was the most expensive telescope ever built but had several hiccups before it started working and doing photography. It was primarily possible by a heroic deed performed by the NASA team and their Astronaut Professor Jeff Hoffman, who carried out a fine repair work that took five days, reaching there in another space shuttle with telescope still in the space. Hubble telescope captured the final moments of a star's life, its explosion and release of gases, *cosmic dust interstellar nurseries* of newborn stars and dart pillars of *cosmic dust* millions and millions of miles long ready to span a new generation of stars and planets. The telescope also captured the tapestry of distinct galaxies in seemingly an empty patch of space, formed millions and millions of years ago while it also noticed some of the first galaxies created after the Big Bang. A large number of galaxies could be seen not known before. Hubble telescope testified what had been just hypothesised. The universe was still in the making. At one point the universe entered into what is generally called a super-creative phase. Then came a situation when the giant stars called supernovae exploded to create more elements which were not created till then. It was roughly a period of 500 million years that our universe had lived through and there were 13 billion years more to go before the appearance of humans on the face of the earth. The telescope saw the explosion of stars and supernovae, who are the creators of new elements, the array that we see today. It is therefore very pertinent to state that without the birth and death both resulting from the explosion, our planet would be dull, possibly might not have forged such a variety and shockingly we might not have been around. The making, however, continues unabated. The fact, as stated by scientists, is that our lives in a real sense depend on the lives and deaths of stars.

The images snapped by Hubble telescope elaborated the aftermaths of the explosion of stars. It must have looked like some incredible and heavenly fireworks over an enormous span of time. Nebulae, giant clouds of debris thrown off by exploding stars, comprised of new atoms like gold, silver, zinc and lead. During its phenomenal services rendered for about 30 years, in addition to millions of pictures, Hubble has observed distant stars and measured the age of our universe. It has shed light on various mysteries of our universe. After such a successful service, Hubble has peaked in terms of how far back it is able to observe. One of Hubble's astounding observation came up in 2016, when it captured the image of galaxy GN-z11 found in the constellation Ursa Major, It was probably the most distant and currently the oldest galaxy observed by Hubble telescope. This galaxy is 32 billion light-years away. As rightly propounded by Edwin Hubble the universe is still expanding. Due to this expansion, the light from the high-redshift galaxy observed by Hubble telescope was 13.4 billion years in the past, just about 400 million years after the Big Bang. This is the maximum Hubble can observe and not beyond it. In fact, Hubble's capabilities are limited to the range of certain wavelengths of light. Galaxies are on the move; when they travel from one point to another in the space the wavelengths of its light are stretched due to constant expansion of space. By the time the light reaches Hubble it is stretched to a wavelength outside the viewing range of Hubble. It cannot observe anything stretched to near-infrared range. The relentless quest of the human mind grows and the paragon of science enriches by leaps and bounds. NASA is likely to substitute Hubble telescope with a more versatile version of space observatories, the James Webb Space Telescope (JWST) in the near future. James Webb once in action, will virtually substitute the Hubble which successfully added to the vibrant treasure of humans' space knowledge for over three decades. This observatory will be able to see what the universe looked like around a quarter of a billion years or possibly even 100 million years, after the Big Bang. Presumably, it was this period when the first generation of stars and galaxies started to form. The difference in the size of mirrors in Hubble and James Webb is conspicuous. Hubble had a 2.4 Metre-diameter mirror while James Webb will have a 6.5 metre-diameter mirror. The number of wavelengths that can fit in a mirror depends on the size of the mirror. A larger mirror allows a higher resolution. Interestingly, the James Webb with its capabilities can observe a small coin from 40 kilometres, a fact that speaks volumes of its breath-taking capabilities.

This may not be out of context to elucidate a rather more modern concept on the matter. However, this does not impart anything to demean the work done by the illustrious astronomers in the past. In the year 1964 Scottish physicist Peter Higgs and others laid out a theory on the advent of everything in the universe and thus the universe itself. It is the physical proof of an invisible, omnipresent field that provided mass to all matter right after the Big Bang. This forced the particles to coalesce into stars, planets and others in the universe. A model in quantum physics comprises a hypothetical, ubiquitous quantum field which is thought to give mass to all the particles in existence. This indeed explains why at all the particles have any mass. This field was named as Higgs field and the particle associated with it is called *Higgs Boson*. The existence of Boson is not yet proved, yet if it is assumed that the *Higgs Boson* does exist, its mass can be inferred on the basis of the effect it would have on the properties of other particles and fields. The *Higgs Boson*, later on, came to be known as *'God particle'* also, probably due to the pious belief that the God is omnipresent, is pervasive and runs the 'entire' show in the universe. In layman's language, it is the *Higgs Boson* which is responsible for the existence of everything—be it of any shape, size and function, if just present in the universe.

The name came into existence in 1993 from a book on the topic. In order to fill in the gap, particularly for those new to the subject, one resolutely needs to comprehend what 'Boson' has to do so with Higgs particle. The name Boson was coined by Paul Adrien Maurice Dirac, an English theoretical physicist. Boson is one of the two classes of particles, the other being 'fermions' and is doggedly bonded to commemorate the contribution of Satyendra Nath Bose, an Indian Physicist and Professor at University of Calcutta and Dhaka, who developed Bose-Einstein statistics which theorises the characteristics of elementary particles. The *Higgs Boson* is an elusive particle since its inception. In fact, no experiment has observed the *Higgs Boson* to confirm the theory. *CERN*, a multination organisation after conducting experiments like A Toroidal LHC ApparatuS CMS combine *(ATLAS)*, the general particle detector on Large Hadron Collider *(LHC)* claimed to have observed a new particle in the mass region. This is consistent with the Higgs particle. ATLAS is one of the four experiments at LHC at *CERN*. *ATLAS* is aimed to answer the fundamental questions like what the basic building blocks of the matter are and what the fundamental forces of nature are. It also aims to understand if there is an underlying symmetry to our universe. With the colossal magnitude of the cost and the utility of research, *CERN*, the European Organisation for Nuclear Research and an acronym for *Conseil Europeen por la Recherche Nucleaire (CERN)* is the most appropriate establishment to undertake such experiments. The reference is probably incomplete if the name of Francois Baron Englert, the Belgian theoretical physicist who shared the 2013 Noble Prize in Physics jointly with Peter Higgs for their work on "theoretical discovery of a mechanism that contributes to our understanding of the origin of mass of subatomic particles", which recently was confirmed by the *CERN's LHC*.

Scientists have also propounded theories about the death of the universe. One hypothesis propounds when the universe will run out of steam and stop expanding and as a result every star, planet, galaxy and the supernovae, every atom will start collapsing, culminating in a super-dense pinpoint, called the "Big crunch". Scientists like Saul Perlmutter have studied if the universe was still expanding. By focusing to study the explosion of Type 1a supernovae and by comparing the dates and positions of supernovae, stretched over space and time, it is calculated if the expansion of the universe is slowing down. Interestingly it is observed that it is not slowing down at all. So another theory, diagonally opposite to the previous one was propounded. The universe, everything included may rip apart due to the expansion to the ultimate limit. Ultimately, it is presumed that the universe will be dark, empty and probably lifeless.

After the Big Bang and passage of almost 9 billion years, the universe had all the necessities for life now in place. The universe was not artless anymore; it has now grown into a vast complex of billions of galaxies and innumerable stars. Silently in a corner in Milky Way, an enormous amount of dust and gas started accumulating which comprised the rich debris left over from one of the supernovae. When this debris reached the critical mass, it began to burn brightly. Consequently, a star was born, the sun, our own star. What was left over, formed a disc of swirling debris that orbits around the new star. The gas and debris that make up this disc collide and is pulled together by gravity. More and more dust and gas get supplemented to make it bigger and bigger and thus the planets are formed. Our earth is one of them. In the following 500 million years, the planet generated a canopy of gas around it, the atmosphere.

The solar system consists of a star, such as the sun and the objects like moons, asteroids, comets, rocks, dust and meteoroids affected by its gravity. The solar system we refer to by this very name and the one known to mankind is constituted by the sun together with its planetary system, which also includes the earth. There are eight planets, in addition to the sun in the solar system. They differ a great deal from one another. There are inner planets Mercury and Venus, closest to the sun; mysterious Mars; accompanied by

Jupiter, Saturn, Uranus, Neptune and the only planet that harbours life in an unimaginable variety of forms, our Earth. Mercury, Venus, Mars and Earth are terrestrial planets whereas Jupiter, Saturn, Uranus and Neptune are Jovian planets which have no solid surface and are gas giants.

What we know of the solar system, elucidates that the solar system is the one of which our earth is a part. Our solar system is just one specific planetary system, a star with planets orbiting around it. There may be more planetary systems present outside our solar system. There are about 200 billion stars in our own galaxy including the sun and earth. As regards the moon, it is the only natural satellite of our earth. Some known planets outside our galaxy are *HD 40307*, with eight times greater mass and gravitational pull. Another is Kepler-16b. It is fascinating to understand that this one is said to orbit two stars. There is *Trappist-1*, a planetary system that comprises as many as seven planets similar to the size of our earth. Although the universe encompasses both the solar system and galaxies, the actual size of the solar system is extremely hard for a human brain to truly comprehend in terms of scale. Life on earth largely owes its existence to the sun. Sun is crucial to virtually what we come across every day during the entire life span on the planet earth. The solar system has the sun at the centre with the earth orbiting it. The heart of the solar system is occupied by sun, where it is by far the biggest object, holding 99.8% of the solar system's mass and is about 109 times the diameter of the earth. The size of the sun can be understood by the prognosis that a million earths could be lodged in it. Due to its extreme gravitational force, all the planets started orbiting around it. The notion of a heliocentric solar system, with the sun at the centre, was possibly first suggested in the Vedic literature of ancient India, which often refers to the sun as the "centre of spheres". Later on, this came to be known as heliocentric. Some interpret Aryabhatta's writings in Āryabhaṭīya as implicitly heliocentric. A prominent school of western thinkers give credit to

Polish mathematician and astronomer Nicholas Copernicus, who in his book *"On the Revolution of the Heavenly Spheres"* propounded that the planets orbited around the sun and not Earth. In terms of size and distance respectively, if the sun is as big as a tennis ball, the earth would be the size of a sand grain, located almost 8 metres or 26 feet apart.

Moon is a natural satellite of our planet and fifth-largest among the moons in the solar system. Our planet's wobble is stabilised by the presence of the moon which stabilises our climate. It has a very thin atmosphere, called the *Exosphere* and is located about 385,000 kilometres away. Some planets in our solar system have one or more moons. The moon that we see somersaulting every clear night, mesmerising the humans is a natural satellite of our earth. Mars has two, and Jupiter has as many as 50 known moons, with 17 more awaiting confirmations. Saturn has 53 known and nine awaiting confirmation. Uranus has 27 whereas Neptune has 13. We comprehend more of our moon in our strides ahead.

The evolutionary history of life on earth is an extremely ambiguous spectacle which has been baffling the anthropologists, the geologists, biologists and has given rise to various schools of hypotheses. It is truly intriguing, obviously for many reasons. Our recent ancestors have been mute witnesses to the extinction of many species in the wild and growth of many others. Just after its inception, the planet earth had a very high temperature. Majority of the surface of the earth was covered by hot boiling lava, the molten rock material. Following 600 million years planet earth gradually cooled, during which innumerable meteoroids and asteroids kept on hitting the earth surface. One school of scientists believe that these meteoroids and asteroids brought minerals and water along with them. The set of scientific work, in the field of paleoenvironment, palaeontology, the ongoing process of global warming and climate change have altogether provided enough support to interpret that the life did not ascend all at once. Life emerged and became extinct many times on the planet. It did not occur all at once, undertook the journey to acquire the present form and its inhabitants. However, the last time it took birth, maybe 4.5 billion years ago, has not perished since then and is thriving. By any magnitude of logic, the evolutionary approach looks more reasonable which, in fact, is supported by adequate pieces of evidence also.

# Chapter 2

# Mother Earth and the Antiquity of Life on the Planet

Of eight planets in the solar system, Earth is fifth-largest, is densest and is the largest of four terrestrial planets. Earth formed approximately 4.54 billion years ago, and life appeared on its surface within its first billion years. This layer along with Earth's magnetic field shields harmful solar radiation. Life was limited to oceans in the beginning. The interior structure of the Earth is layered in spherical shells, like an onion. The three main layers are the core, the mantle and the crust. The core is constituted by inner and outer cores; the centre is 6378 km. ending at 5100 km. from the surface, followed by the outer core which ends at 2900 km. from the surface. The inner core is solid, has the radius of about 1278 km. while the outer core is liquid and is about 2200 km. thick. The layer above the core is the mantle. It begins about 10 km. (6 miles) below the oceanic crust and about 30 km. (19 miles) below the continental crust. Moving outwards from the centre, the mantle is followed by the crust. It is the earth's hard outer shell, the surface on which we are living. The oceanic crust is about 6-11 km. (4-7 miles) thick and consists of volcanic rocks like *basalt*. Continental crust is thicker than the oceanic crust and is mainly made up of igneous rocks like granite.

Earth's biosphere conspicuously altered the basic physical settings that facilitated the propagation of organisms and the formation of the *ozone layer*. The earth is surrounded by colourless, odourless, tasteless 'sea' of gases, water and fine dust which is called the atmosphere. Due to the gravitational force, the earth interacts with objects in space, especially the sun and moon. Moon, the only natural satellite of earth began orbiting about 4.54 billion years ago. The gravitational interaction between earth and moon initiates the ocean tides on the earth, reduces the speed of rotation and stabilises the axial tilt.

One of the prime reasons why life on the earth is possible and thrives is due to the presence of the atmosphere around. The sun resolutely imparts compassion towards the earth providing the most essentials for life to survive and thrive in a variety of forms. Planet earth is home to millions of species of life, including humans. Both the resources occurring naturally on the planet and the products of the biosphere contribute to the resource reserves that are extensively used to support the global human and animal population.

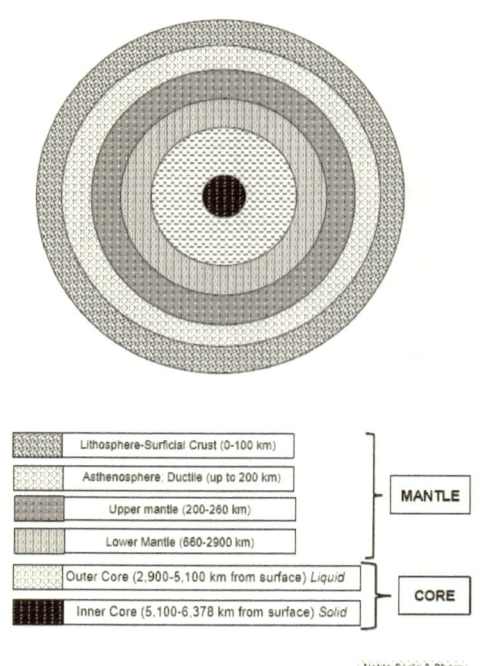

Layers in the Generalised Interior of the Earth

In spite of lengthy monologue often presented for the benefit of persuading an in-depth perception about the intricate problem as to how and when the universe, sun, moon, planets and earth came to exist, we cannot ratify any of the ideas with true conviction. Most of the hypotheses are based on presumptions, though supported by strong, rational and coherent perspectives. Virtually the same stands true for the intricacy related to the origin of life on earth. How the life originated is truly the most fundamental, yet baffling and at the same time the least understood biological problem for biologists—equally for zoologists, botanists, anthropologists and geologists. One of the following four categories of hypotheses cover up the diversity of notions put forward to explain the origin of life:

1. Life on earth appeared spontaneously, particularly the simpler forms and readily arose from non-living matters in short periods;

2. The life resulted spontaneously from a supernatural event, which is irretrievably beyond the descriptive capability of any of the discipline of science;

3. Life came on earth along with the matter, thus these two are coeternal. It came on earth either simultaneously with its origin or shortly thereafter; &

4. Life arose on the young earth by a series of progressive chemical reactions, which may have been likely or may have resulted from one or more highly improbable chemical events.

Apparently, none of the four can be totally endorsed, yet they have provided a platform for further brainstorming and provoked the inquisitiveness to the scientific bend of mind.

This is, however, a more logical proposition to go about what appears to be closer to the facts and satisfies the analytical faculty of the human brain. What has been incorporated ahead, done after torrid consultation of the related literature, is based on observations, being fully supported by most reliable contextual scientific work performed by various bright, competent workers and explorers who deserve a rousing applause. The depiction ahead exhibits this.

We briefly but lucidly reproduce ahead, tracing events from the formation of hydrogen in the solar system, to the stage of evolution of modern *Homo sapiens*.

Approximately 4.54 billion years ago, the solar system also called the solar nebula, was in prodigious mayhem, had only evolved the hydrogen and helium atoms but without electrons which later fused to achieve wholeness when the immense heat and speed had subsided and the universe was comparatively calm. A molecular cloud, a vast build-up of dust and gases got accumulated in one of the corners of the universe. It covered the space hundreds of light-years in size. This cloud contracted under its own gravity and from the central part—more hot and dense, our sun was born. The leftover part of the cloud formed a swirling disc called the solar nebula. The disc comprised Mercury, Venus, Earth, Mars, Jupiter, Saturn, Uranus and Neptune. They started orbiting around the sun. Thus we may enunciate that earth formed by accretion from the solar nebula, approximately 4.54 billion years ago. The earth was just not hot but was virtually a ball of fire. The primordial atmosphere presumably formed by volcanic actions around 3900 million or 3.9 billion years ago. The atmosphere then was composed of gases like carbon dioxide, methane, ammonia, nitrogen and water vapours. All cumulatively triggered the gradual cooling of the earth. Consequently, the condensation of water vapours started to form water resulting in rains and accumulation of water in many forms. The most redeeming feature was the absence of oxygen gas in the atmosphere. This was the feature which in fact, delayed the advent of life on our planet. Earth's surface which was molten and soft in the beginning had also started solidifying. The revolution of the earth on its axis was much faster than what it is today. Thus the day then was around 15-16 hours. The meteoroids and asteroids falling from space penetrated through the molten surface of the earth due to shear force and many of them pierced deep into it. This bombardment had formed huge pits that now became suitable venues for water to get accumulated. Majority of basic needs for life were now available on the earth. This gave rise to life on its surface in the form of unicellular organisms. This happened within the first billion years of its formation. These organisms were mainly bacteria that would make their food with sun rays. Then about 2.28 billion years ago, oxygen was beginning to be formed by these organisms, which also heralded the advent of photosynthesis on the earth.

This is what the scientist call *Great Oxygen Event.* This truly was an immensely vital phase in the evolutionary process. For the next tens of millions of years, this process continued and the atmosphere of the earth became abundantly filled with oxygen. This account has received concrete sustenance recently in the form of remarkable discoveries. The earliest evidence of life is traced from biogenic carbon signatures and the fossils of *Stromatolites*, retrieved from metasedimentary rocks of Greenland which are dated about 3.7 billion years old. This came up in 2015. Later in March 2017 possible material known as the "remains of biotic life" have been found in 4.1 billion-year-old rocks in Australia.

Around 2200 million or 2.2 billion years ago the inception of organisms laden with mitochondria followed, which made the use of oxygen in respiration. Later on, about 2000 million years ago a large meteoroid crashed on the earth's surface to form a large pit of about 325 kilometres, present even on this day in South Africa, known as Bredford Dome. Again about 200 million years later another mammoth meteoroid crashed on the earth creating a depression of about 250 kilometres, present in Ontario, Canada. This process of destruction and construction continued during the next several million years. The extinction record in the history of the earth is often taken as events of inordinate value. Since its inception and post the Big Bang, our universe has been the taciturn witness of mainly five great extinctions. These catastrophes had left their prominent signatures on the erstwhile constitutions and the inmates of the universe. Their occurrence was also instrumental in framing the future shape and design to a considerable extent. They are named after the geological period and/or epochs of their respective occurrence. These extinctions are also referred to as mass extinction or biotic crisis. These events left a very powerful impact resulting in a prominent decrease in the biodiversity.

## End Ordovician or Ordovician-Silurian Extinction

The period which depicts the end Ordovician period and the beginning of Silurian period marks the incidence of two events probably in succession. In terms of the geological calendar, the period may have been 450–440 million years ago. These events wiped off about 27 to 28% of all families, 57% of all genera and 60%-70% of all the species. In terms of devastation and the extinction of the genera, these events are considered as the second largest in earth's history.

Interestingly, multiple causes are thought to have created this havoc. In addition to massive tectonic activities, resulting in large scale volcanism, the formation of new mountains like the Appalachians, movement of tectonic plates of the supercontinent Gondwana, collision and folding of tectonic plates against each other, were certain other astonishing factors which are assigned to cause this extinction. The land was barren during that period and life had already originated in the ocean. Scientists consider the terrestrial part as devoid of any life for some time. But some thought that clinging to the rocky layer and soil cover was a micro-thin layer of minute bacteria, representing terrestrial life on the planet which it is thought were *blue-green algae* or the *cyanobacteria*, also known by the collective name of *Stromatolites* or some primitive algae. They were there since Precambrian times. These tiny terrestrial plants are considered to have caused the environmental catastrophe which occurred almost on the border of the Ordovician and Silurian periods. The fossils of spores of plants found in Saudi Arabia and the Czech Republic, which were dated to 462 million years ago and later on in Argentina which were dated to about 470 million years ago account for discoveries in support of this idea. The thick covering on these spores proves them to be similar to spores from terrestrial plants today which act as their protective shield to successfully confront the environmental stressors like wind and surface run-off of water. On the contrary, the aquatic plants do not have such covering, as they do not face such hardship. However, the plants are thought to have reached the land even earlier, an interpretation based on the dates arrived at by the method called Molecular Clock. This method gives the date of 511 million years ago, right in the middle of the Cambrian period. Argentinian fossils do not confirm the spores on these plants but clearly establish them as non-vascular plants or the ones that do not have roots. In due course of time, through natural selection, they adapted to the terrestrial environment and developed hard walls and thick waxy coating called the cuticle. The material clinging to the rocks is cryptogrammic cover, comprising algae, moss and lichens, etc. The rock clingers break the rocks releasing the minerals like phosphorous, potassium and iron. By recreating the situation in a laboratory, the Palaeontologists have shown that in case of rocks without moss, 60 times more phosphorous is released which is carried to the ocean by surface run-off. Since phosphorous is a rich nutrient which plants need for growth, such an explosion of phosphorous into the ocean would have caused an explosion of marine vegetation, creating an enormous algal bloom. They eventually die and are broken down by bacteria. This process consumed a lot of oxygen in the water, the oceans then first become hypoxic and later on anoxic. Marine animals must not have had any oxygen to breathe, succumbing to anoxia. The hypoxic or anoxic oceans can also cool the climate. Carbon present also needed oxygen to bind with it to form carbon-di-oxide. This could not have happened to a very large extent, due to hypoxic or anoxic atmosphere. The carbon got deposited as black shale which may be seen even today as mammoth deposits of black shale in China and Africa dating to late Ordovician. So a much cooler climate and oxygen-poor oceans could also possibly have been the factors behind the extinction of ocean life.

## Late Devonian or Devonian-Carboniferous Extinction

A prolonged series of extinctions took place during the period which marks the junction of Devonian and Carboniferous periods—about 375-360 million years ago. These events eliminated about 19% of all families, 50% of all genera and 70% of all the species. In terms of devastation and the extinction of the genera, these events are considered a significant event in earth's history. Literally, the earth was pulsating for a period of about 20 million years which is evident from the assemblage of index fossils. Index fossils are the ones which survived during a particular geological period/epoch and became extinct during the same period/epoch.

## End Permian Extinction

The transition of Permian and Triassic periods, about 252 million years ago probably witnessed the biggest extinction in earth's history, which eliminated almost 57% of all families, 83% of marine and 90 to 94% of all genera, of which majority was marine, while 70% were terrestrial including insects. Also known as the *'Great dying'* and is very significant in the evolutionary history of the earth, it completely eliminated the marine Arthropods and Trilobites, considered to be highly adaptive and successful species. It also ended the primacy of mammal-like reptiles. It took the next 30 million years to recoup the loss of vertebrates, consequently paving the way for *archosaurs* which became dominant. Broadly, the whole late Permian was a difficult time, particularly for the marine life. This event of extinction is considered to be the largest in the history of the earth.

## End Triassic or Triassic- Jurassic Extinction Event

At the end of the Triassic period and the beginning of the Jurassic period, about 201 million years ago, an event of extinction took place. This eradicated about 23% of all families, 48% of all genera, 20% of marine families, 55% of marine genera and about 74% of all species, together with most of the non-dinosaurian *archosaurs,* large amphibians were eliminated from the planet. This gave the dinosaurs virtually no rivalry on the land whereas the marine life was dominated by non-archosaurians. A lineage of large amphibians, however, survived in Australian sub-continent. They were *Koolasuchus*, a genus of brachiopod temnospondyl, considered to have lived till Cretaceous period but became extinct about 120 million years ago.

## Cretaceous-Paleogene Extinction or End Cretaceous also Known as K-pg, Formerly K-T Extinction

This event, now officially called K-pg extinction occurred about 65 to 66 million years ago. It caused the extinction of about 16% of all families, 50% of all genera and 75% of all species. As regards marine life, all the *ammonites, mosasaurs* and P*lesiosaurs* vanished. About 33% of the immobile or sessile animals got eliminated. All the land-dwelling dinosaurs, non-avian in general, were eliminated during that period which included the most diverse group of the land-dwelling giant dinosaurs like *Giganotosaurus* and the famous *Tyrannosaurus*, popularly called *T. Rex* belonging to Theropods clade. The dinosaurs lived on the earth for over 165 million years but all were eliminated during the Cretaceous period. This event was severe with a significant amount of variability in the rate of extinction between and among the different clades, mammals and birds. There is an ongoing process of extinction resulting from the activities of the human world: what we see today but are unable to comprehend due to extremely slow pace of the change which probably the human brain cannot view. Some name it as Holocene extinction in which human activities also contribute significantly. This, therefore, must not be clubbed with five great extinctions whose cause was utterly natural. The process is on and the assessment by Intergovernmental Science-Policy Platform on Biodiversity and Ecosystem Services *(IPBES)* asserts that there are about 8 million species estimated to be present on the earth and about 1 million plants and animal species are threatened with extinction.

| GREAT EXTINCTIONS IN THE DEVELOPMENT OF LIFE ON EARTH | | | | |
|---|---|---|---|---|
| Sr. No. | Period | Years | Percentage (%) of Extinction | Anticipated Primary Causes |
| 1. | Ordovician-Silurian Extinction | 450-440 Ma | 27% of all families, 57% of all genera and 60 to 70 % Of all species | According to a very recent discovery Global warming due to volcanism, warming, anoxia and not due to cooling and glaciation |
| 2. | Devonian-Carboniferous Extinction | 375-360 Ma | About 19% of all families, 50% of all genera and at least 70% of all species | Tectonic activity, climate & sea-level fluctuations, volcanism and collision of earth with huge cosmic bodies |
| 3. | End Permian-Permian-Triassic Extinction | 299-252 Ma | 57% of all families, 83% of Marine and 94% of all genera and 70% of terrestrial fauna became extinct | Massive volcanic eruptions of more than 4 million cubic Kilometres of lava over Siberian Traps, Siberia Russia |
| 4. | End Triassic-Jurassic Extinction | 201-252 Ma | 48% of all genera 20% of Marine families, 55% of marine genera and 74% all species | Most scientists agree to a massive eruption of lava and $CO_2$ from Central Atlantic Magmatic Province (CAMP) |

*Continued...*

| 5. | K-pg or K-T extinction | 65 Ma | About 16 % of all families, 50% of all genera and 75% of all species | Impact of a large comet or Asteroid 10 to 15 km. wide, devastated global environment |

The earth is located at the far end of the Milky Way galaxy where about 5 million years ago there was a large cloud of gas, dust and particles of sand, etc. from which it was born. Different schools of scientists have lately agreed more or less unanimously that the planet we now sit on was formed from this cloud. They call it a molecular cloud. The large molecular clouds rotated slowly in small circles but on shrinking, their rotation became much faster in spite of the contents remaining the same. After birth, the evolution of earth is best explained by an easily conceivable storyline spanning over a period of 12 hours on the clock. The incredible journey spreads from its birth to the present. The earth is born, as it spun off from the rich debris left over from one of the supernovae and settled at the far end of Milky Way galaxy. It is 12 midnight, the clock starts ticking and the earth's journey to become a planet begins.

Once the bunches of particles in the molecular cloud grew into objects just half-a-mile in diameter, their mass was great enough for the gravitational pull to attract material from the surrounding discs. Circling around the sun it continued to suck the material until nothing was left to consume. In the inner circle of the solar system, these bunches grew into 22 planets. The earth had, by then, lived for about 3 million years since its birth, on the clock it was less than 30 seconds, post the beginning. This was followed by an incredibly violent phase. These planets orbited the sun. As a result of their gravity affecting each other, they began to collide. With each collision two planets combined. Over the period, collisions reduced the number of planets in the inner solar system to just a few, including Venus, Mercury, Mars and Earth. Astronomers believe that the earth then was about 30 million years old and our clock had moved about 5 minutes.

The earth then was hot with its temperature at more than $4700^0$ C. While the earth started to cool, it faced a serious threat by a hurricane of charged particles from the sun. It could have wiped out the chances of any life on the earth, had it struck. Even a light breeze from the sun travels at about 200 miles per second while the temperature touches $4500^0$ C. The storms on the sun affect us here on the earth. Due to the powerful explosions in the sun, energetic particles are released, creating what is known as solar winds. These particles are a form of radiation, travelling at a speed of 1 million miles an hour. They can have a devastating effect on the planets. In due course of time, the earth got enveloped in a thin layer of gases, the atmosphere which protects it from the ill-effects and extreme temperature of the outer space. The solar winds are, however, so intense that they can rip off the atmosphere. We can learn this from what these solar winds have done to planet Mars. These devastating solar winds have rendered Mars completely devoid of its atmosphere, left it without any liquid water and too little air to breath. This could have happened to our earth also, yet there was something that stood in the way of destruction. In the beginning, the temperature on the earth rose so high due to the heat generated by material hitting it from the outer space that even the rocks remained in the molten stage. As a result, lighter elements floated in the upper portion while heavier ones sank to the centre including iron. The earth thus became endowed with a molten core. It was this strong molten core that protected the planet from adverse and often deadly effects of the sun and so it does even today. The spinning core creates a magnetic field that surrounds the earth. This happens exactly similar to what happens in a dynamo. This magnetic field gave our earth its north and south poles. The field reaches far into the space to form what is called as *Magnetosphere* which protects us from solar winds. Majority of the particles in solar winds are prevented by the *Magnetosphere* and a few that penetrate are deflected towards the poles. Although the atmosphere is being depleted by solar winds even today, the pace of depletion is too slow to reckon with. According to one estimate, it might take many times the life of the sun. The magnetic field played an indispensable role in the evolution of earth and our survival on the planet, probably without which we would not have any air to breathe at all. The earth was cooling down but it still had no oxygen and water. Barely 6 minutes have passed out of the total of 12 hours on our clock. Still, the temperature was tremendous—about $1100^0$C, impossible for life to sustain. Presumably, there was a likely collision of planet earth with another planet of enormous size—a collision so vast that it could have melted the whole planet.

The moon's presence in the space is the only proof which shows that such an event occurred. Regarding the formation of the moon, many theories were propounded but a plausible explanation could not be found over centuries. Some thought that the moon was formed by the early earth which was spinning so fast that it threw some material into space which is the moon. Some others thought that it was just a passing planet which got captured by the gravity of earth. Neither anyone really knew nor could anything be explained. In the year 1963, the United States launched the Apollo programme. One of the missions was to find out how the moon was formed. They made six trips to the moon till the beginning of the seventies. The astronauts collected 385.5 kilograms of moon rocks to conduct detailed studies back home. The rocks collected showed an extreme state of desiccation as if they were heated to a high temperature. This baffled scientists. In the decade of nineties, the planetary scientist Robin M Canup of Southwest Research Institute, Boulder, Colorado, USA, the propagator of *Giant Impact Hypothesis* to explain the formation of planets, argued that moon and Earth formed in a series of steps that started with the collision of two planetary bodies. She hypothesised explaining through computer simulation by creating a model that an incredibly large planet about half the size of the earth, named Theia, racing towards the earth at seven miles per second, hit the earth obliquely, almost at an angle of $45^0$. The impact would have been an incredibly energetic event with enough energy to completely melt the whole of the earth, vapourising a significant portion of the rocks had the collision been direct. But it was just a glancing impact which smashed some material into the space. Much of this debris stayed in the orbit as a disc of dust and gas. In due course, this clump of circling material became large enough for its gravity to suck in more material from the disc. This was the moon that has been fascinating the humans over centuries. Moon is our closest neighbour and keeps our earth stable in its orbit. It is our permanent natural satellite. The largest of the moons is Ganymede, a satellite of Jupiter, about 0.41 times the earth's diameter. Titan, the second in size, is a satellite of Saturn. Third in the row is Callisto while Io is fourth, both satellites of Jupiter. There are as many as 200 moons in our solar system, orbiting other planets. The moon that we have known for times immemorial, is the fifth-largest. It is due to our moon that we have 24 hours making a day on earth and we see stable seasons on earth. There are other small objects orbiting our earth, called a Neos, the "near-earth-objects". When the solar system was 30 million years old, the moon was formed, following the formation of the earth, about 4.5 billion years ago. It formed by the collision of Theia and the earth; thus it is a part of earth. Many mythological beliefs across the world recognise the status of the moon as a revered deity. Hindus recognise the moon as "Chandradev", the father of "Budh", the planet Mercury. Due to the orbital position of the moon around the earth, there is a 15 days' period, known as 'Poornima', the 'Shukla Paksh', or the waxing moon, the bright lunar fortnight, the rest of the 15 days' period is known as 'Amavasya', the 'Krishna Paksh' or the dark lunar fortnight or waning moon in a month as described in Hindu mythos.

It was 50 million years since the earth began to form and the clock on which 12 hours represent the whole of earth's history, only 8 minutes have passed. The time is 8 minutes past 12 midnight. The ground remained molten at this stage and the moon was 15 times closer than it is today. As described by scientist Robin M Canup, the full moon night would have been breath-taking at that time with moon 15 times larger. The collision with the planet and the formation of the moon were key events for creating and making the earth fit for life. The tilt also probably resulted from the collision and it is now established that this tilt gives us the seasons and the annual cycle of life. Creation of moon triggered the incidence of tides in the ocean. They must have been much stronger when the moon was closer to the earth than what we see today as the moon slowly drifted away and is now located far away from the earth. The scene refers to the situation after which the oceans and water had appeared on the earth.

Another redeeming feature that existed in all its fairness is that there was no oxygen and there was no water either on the earth. Some scientists believe if there was any on the earth, it was very little of these essentials. Oceans cover about 71 to 72% of the earth's land area. There are five oceans spreading across the planet. Out of earth's total land area of 510072000 square kilometres, oceans, five in number, occupy about 361132000 square kilometres, a whopping 71-72%. The remaining 148940000 square kilometres is the landmass. There is about 326 million trillion gallons of water on the earth at present. Oceans constitute its major chunk. Where and how did they appear when there were none initially? The planet was so hot that there could not have been any liquid water on it at all. The region from the sun to the asteroid belt was bone-dry. The source of water reserves, nearest to the earth was about 160 million miles away in the asteroid belt. How it came to earth and that too in such colossal quantity is a mystery that scientists are working on to solve. In the year 2005, NASA launched an unusual spacecraft, named *Deep Impact*. Its objective was to fly to reach the comet named *Temple-1* and launch a 365 Kilograms copper impactor on to its surface to reveal the internal structure of the comet, particularly if it had water and ice. Deep Impact remained in the space for nearly nine years. For a very long time, it was thought that comets contained water in the form of ice and that they only had brought it to the earth. Deep Impact directly hit the object. The debris on analysis showed that the comets had significant amounts of water. Vapour trails from comets were studied using a radio telescope to perceive if the water in comets matched with the water in oceans. The comet theory, though not dead, discouraged its proponents. In the year 2000 Canada witnessed the fall of a meteorite from the space on the frozen Taggish lake in British Columbia. The meteorite pieces were collected from Lake Tagish and studied by Dr. Michael Zolensky, a scientist in NASA and concluded that it was composed mainly of clay minerals. The molecules of clay minerals contained water and about 20% of it was water. It was a substantially large quantity. This means, on hitting the ground on earth the meteorites become very cool. On tracing the Tagish Lake meteorite's orbit back in the space, the scientists found that it came from the outer reaches of the asteroid belt. The amount of water entirely depends on the distance of the meteorite location to that of the sun. If the asteroids were the source of water and oceans on our earth, the conclusion baffles as to how such a large number of asteroids left their normal path to head towards the earth to crash. Jupiter, the largest planet in the solar system lies just behind the asteroid belt. In the distant past, Jupiter's gravity disrupted the circular path of asteroids and charted an elliptical orbit for them. Thus their collision with earth became inevitable. The earth was bombarded. As the asteroids crashed, they smashed and water inside them escaped. On repetition of a natural phenomenon of such a magnitude, oceans were created that we see today. This process did not take long. A Geologist Dr. Stephen James Mojzsis in his laboratory, at the University of Colorado at Boulder, Colorado extracted zircon crystals from some of the oldest rocks. With the use of Ion Microprobe, he measured the composition of oxygen isotopes inside zircon crystals. It was observed that these crystals formed a particular type of rocks which could have existed only if there was water on the earth. The hypothesis that the asteroids brought water on the earth, further leads to the perception that our oceans came from space was now supported substantially by the science. The asteroids came crashing from a torrential shower of meteorites on the earth. Dr. Stephen Mojzsis answered many ambiguous questions, who in the course of his experiment, also estimated the age of zircon. "These zircons tell us that they melted from an earlier rock that had been on the Earth's surface and interacted with cold water", concluded Mojzsis. A distinctive oxygen isotope signature was encountered and established to be older than the earth's oxygen atmosphere. The isotopic signature clearly indicated that the zircon interacted with cold water on earth's surface about 4.3 million years ago. Thus the conclusion could be drawn that the water arrived on the earth before the earth's oxygen atmosphere.

On our clock, about 25 minutes have gone by since the earth was born at 12 midnight. The *hydrothermal vents* brought iron, due to which the oceans became iron-rich, thus reflecting the green colour. The atmosphere might have been much denser than at present, thus reflecting reddish tinge of colour. The scene must have been fascinating. The planet had also cooled down to about $100^0$ centigrade. Apart from the temperature dropping close to about $100^0$ C, there was another horrendous deficiency. The tempestuous onslaught on the earth from the sky had also subsided about 500 million years after it began to form. The combination of gases in the atmosphere was conspicuous by the striking absence of oxygen; it comprised gases like nitrogen, hydrogen and hazardous methane. No human or animal could have survived even for a minute. The inquisitiveness where this oxygen came from prompted many anthropologists and earth scientists to undertake to resolve the issue. Dr. Martin Van Kranendonk of Geological Survey of Western Australia successfully came up with plausible elucidation. Hamelin Pool Marine Nature Reserve, Shark Bay of west coast, Australia, is a world heritage site endowed with meadows of luxuriously growing seagrass, marine herbivorous mammals, the innocuous *Dugong* and tamed *Dolphins*. This is also one of the two places where living *Stromatolites* are present. These strange but valuable organisms have been growing in Shark Bay since the last Ice Age, about 10000 years ago. *Stromatolites* are primitive organisms and are known to exist in Precambrian era also, though majority of them fossilised now. They occur in two forms—flat bacterial mats, the short pillars of living bacteria and silt. Part of their body is living *blue-green algae* or *cyanobacteria* that make their own food from sunlight and carbon dioxide by photosynthesis and release oxygen in the process. It is these organisms who could be considered the crowning custodians to make our planet earth a place to survive. What made Kranendonk to get convinced is a remarkable work of research. A remote region of Northern Australia, the place called Pilbara, inhabited by aboriginal people, a global biodiversity hotspot for subterranean fauna and rich mineral deposits, particularly iron ore, is extremely important as it also encompasses one of the oldest rock formations on the planet. The fossil *Stromatolites*, older than the Shark Bay *Stromatolites*, are considered to be oldest fossils on the earth. A prelude that the oldest evidence of life on the earth is present in Pilbara, is just exhilarating. The Pilbara fossils may be great great-great-grandfathers of Shark Bay *Stromatolites*. These fossils suggest that they were formed soon after the earth came into existence, even though no one knows how life began. The bacteria that created these fossil *Stromatolites* lived on the earth one billion years after the earth came on. The time on our clock is 20 minutes past three. The fact that these organisms release oxygen and ate away carbon dioxide, is established beyond doubt. One might perceive that they possibly came to exist soon after the earth was born and this ascertains that there should have been sufficient oxygen in the air. But this did not happen as in the next about 1000 million years, there was none. The banded iron formation found occurring extensively in Karijini National Park in Hamersley Ranges of Pilbara region in Western Australia is considered to be the creation of these *Stromatolites*. The age assigned to these rocks is 2500 million years ago and were formed at the bottom of the oceans. The water in the oceans was heavily laden with iron, thus green in colour. These bacteria present in the sea bubbled out oxygen, which reacted with iron to form iron oxide. The rust precipitated and fell on to the bottom of the ocean. These rocks, according to an estimate are thought to conceal over 20 times of oxygen than present in the atmosphere. Their formation was exceedingly sluggish and they took an incredibly long time to form. On our clock on which the whole history is represented as 12 hours, the major events can be summed up as, the planet forms in the first 8 minutes; after 25 minutes it is cool and collected water on its surface; the removal of most of the oxygen from the oceans and depositing it in these rocks took about the next 4.5 hours. Then after 2500 million years ago the oxygen in the atmosphere started to rise and continued for the next 2000 million years until about 500 million years ago it reached a level at which the animals could survive. This is when the life essentials became handy on the earth and humans could also survive. On the clock that we have taken to represent the story of earth's evolution, 11 hours have passed. At 13 minutes past 11 the dinosaurs appeared and at 50 minutes past 11 A.M., a catastrophe, the K-pg extinction took place wiping out the dinosaurs. Finally, at 59 minutes, 41 seconds past 11, that is 19 seconds to 12 noon the first human appeared. Since then just a few seconds have passed on our clock. After the earth attained sufficient oxygen to breath, about 500 million years ago and the time when the first human walked on the planet, there was a continuous saga unfolding to welcome enthralling flora and fauna in our legend ahead. As far as production of oxygen and reduction of carbon dioxide on the planet is concerned, in Eocene Epoch, that is about 49 million years ago, the appearance of Azolla ferns in the Arctic Ocean, made the matters brighter. Azolla is an aquatic fern, a genus of seven species belonging to *Salviniaceae family*, also called mosquito fern, duckweed fern, fairy moss, water fern and sometimes 'super plant'. It multiplies at an amazing rate and fixes atmospheric nitrogen also. It doubles up its biomass in over 45 to 72 hours and also survives on a symbiotic relationship with other plants. The modern world frequently uses this plant as a bio-fertiliser.

Nevertheless, the antiquity of development of life on planet earth in an exact sense of elucidation, remains unresolved, a matter of further deep probing. The life on the planet took off with the advent of unicellular organisms and the evolutionary history of life followed with hiatus. Broadly, the similarities in life functions among all the living and extinct species clearly indicate that they have diverged through the process of evolution from a common ancestor. The journey of evolution has also encountered the phenomena like DNA mutation of three types, videlicet, base substitutions, deletion and insertion. Nevertheless, the chemistry and processes shared by the entire living world remained unchanged. This strongly supports the view that their origin is traceable to the *Last Universal Common Ancestor, "LUCA"*. This thought irrefutably endorses the basic thought of Hindu philosophy which proposes that everything emerges from a superpower, herein named God and perishes to get assimilated in Him, the God, after going through the tedious journey of life. Undeniable truth is that nothing could have happened overnight, except the spontaneous disappearance. Could that have happened without leaving any sign worthy of scientific cognizance? We will come across such partly or completely unresolved mysteries with some advanced civilisations ahead in our narrative.

# Chapter 3

# A Version of the Birth and Evolution of Human Race on Planet Earth

In order to obtain a clear view of the intriguing puzzle of evolution, based on the preserved set of knowledge the yawning past can be divided roughly into two broad segments: history and prehistory. History recorded what the people practised in terms of food, cultural etiquettes and their beliefs. They adopted writing which was one of the most important inventions of that segment of time. However, in some places, this could not take a shape till recently. The oldest writing is not more than 6500 years old. Broadly prehistory was devoid of the art of writing. The record of human culture existed, verbally passed on from generation to generation and was thus maintained. In the field of scientific ventures fossils have played the role of dossier of the erstwhile world. A part of history was also devoid of the art of writing. This part of history together with the one with the art of writing in place is known as historic period. On the contrary, the record of human culture and experience with no signs of writing in existence is the period known as prehistoric. Our information about prehistory is based on the indirect interpretations obtained by scientists from various objects. This family of scientists includes archaeologists, paleoanthropologists, geologists, palaeontologists, cultural anthropologists, physicists, zoologists, botanists, historians and dedicated amateurs. Their work is difficult, yet it leads to various conclusions that are valuable in advancing the understanding of the past. They adopted to convey their thoughts and findings in written form by using some language—initially pictographic and later in scripted form.

Before moving further in evolution lineage, inching closer to the modern human, it is all the more imperative to comprehend the classification of periods. The development of the size of the brain and similar traits functioned as a hallmark criterion for advancement made by various species in the human lineage. In fact, this led the human species to become hunters and gatherers from scavengers and gradually to qualify for using stone tools. Similarly, another trait was to make use of and control the fire. All this happened in due course and varied from species to species. Based on such benchmarks, the scientists have classified the period from 3 million years ago to this day into two major periods—prehistory and history. The period from 3 million years ago to 3000 BCE is classified as prehistory. The end of prehistory at 3000 BCE saw a major achievement; the humans had invented writing.

The living organisms affect the environment they live in—either make it more lively or impair it. Initially, for billions of years, the earth was not suitable for life to evolve. In due course of time, the earth became more and more favourable for life to grow and multiply unlike in the beginning. Evolution is a dynamic process, continues even today and is responsible for the changes—addition and subtraction in fauna, flora and for the generation of diversity. It is noticeably evident in the fossil records and in the modern ecosystem. Evolution is a process by which organisms evolve from earlier, simpler organisms. The effort is to discuss the evolution during pre-human times, together with human prehistory. Fossil records have been of tremendous support in the interpretation of the deep past of life.

British naturalists Charles Darwin (1809-1882) and Alfred Russell Wallace are regarded as the pioneers in introducing the related concepts. Smithsonian established the first fossil animal museum in 1900, also candidly called the hall of extinct monster. As most appropriately stated by Darwin, "From so simple a beginning endless forms most beautiful and most wonderful have been, and are being, evolved." They observed the diverse fauna by undertaking long voyages—Darwin to South America and Galapagos Archipelago Wallace to South America and Southeast Asia. They observed incredible diversity of life and also noticed the slight modification adopted by similar organisms which ideally suited their environments. Darwin and Wallace in a way were pioneers but scholars like Thomas Malthus, Charles Lyell, Georges Cuvier and Jean-Baptiste Lamarck pushed forward the thoughts that helped in imparting more clarity to the riddles of evolution. They propounded that the earth was very old, that species seemed to change and go extinct over time. The individuals competed over limited resources that were available. Aided by these thoughts Darwin and Wallace brought forward the theory of natural selection based on a series of principles. Their ideas addressed the population and not individuals. In a population of living things, natural variation will occur, due to which some members of the population will survive and reproduce more than others. Ones surviving and reproducing, they will pass on their traits to their offsprings. These traits will give advantage in a certain set of environmental conditions which will be passed on more often; as a result, more members of the population will incorporate those traits and ones failing to do so, remain devoid of those traits and will ultimately get depleted. Gradually only certain traits will show up. When this series of events takes place in a species, we call it microevolution. This is how environmental changes are responded to by a single species. The same plan on a broader scale is called macroevolution. Finally, by the accumulation of the traits over long periods new body plan, new species and new patterns of diversity in the tree of life are created. However, the theory of natural selection was unable to explain how the traits were passed on to the offsprings from parents. No one knew about the genetics at that time. Czech monk named Gregor Mendel was engaged in experimenting on peas plants. He figured out that traits did not simply blend together when living things reproduced and that only some were inherited as discrete traits by a different number of offsprings. Around the turn of the 20$^{th}$ century, this theory was rediscovered by biologist Thomas Hunt Morgan who experimented on flies in 1910. He theorised that mutation was a source of variation in living things and on that variation natural selection acted. Beneficial mutations would be retained and passed on, and detrimental ones would eventually get lost. By this time two of the major forces of evolution were recognised—natural selection and mutation. Population genetics was founded by Raynold Aylmer Fisher, John Burdon Sanderson Haldane and Sewall Wright. They concluded that the natural selection acted slowly but uniformly in a large population. Sewall Wright focussed on understanding how evolution worked in smaller and isolated populations. He applied mathematics to study genetics that led to another key idea that we understand as genetic drift. It is the frequency at which certain genes will sometimes change totally by chance, and will randomly have a greater effect in smaller populations. Then in the 1930s, the idea of gene flow came. These three scientists concluded that the natural selection of genes was the most likely explanation for how evolution works.

In 1937 Theodosius Dobzhansky brought together all the evidence from genetics and natural selection to show how evolution by natural selection could produce a new species. It was an enormous conceptual jump from microevolution to macroevolution. From his experiments, Dobzhansky propounded that new species originate and mutations happen naturally in populations creating variations that can stick around if they are beneficial or just be neutral. If populations are isolated, these variations can remain within a single group with new mutations popping up and none would spread to the rest of the species. Over a while, this would make one group genetically distinct from others, potentially causing problems if it tried to interbreed with others. With enough of a gap in time, it would lose the ability to interbreed with another population group. Subsequently, it would become an entirely new species. This is now firmly taken as the framework for the modern synthesis.

In spite of progression in modern synthesis, the framework remains the same for our understanding of how evolution works. More understanding of how genetics works came up in 1953 when the structure and way of working of DNA were unearthed. Realisation emanated that sometimes the DNA is copied incorrectly during replication and the mutation occurs randomly. This explains how bacteria become resistant to antibiotics over time. We also see that *Pezosiren*, an aquatic mammal is related to the more recent *Metaxytherium*. They are separated by a substantially wide gap of 30 million years; they lost their hind limbs and acquired flippers in place of feet. Their closest living relatives are *Manatees* and Dugongs. Reaching here and obtaining the possible explanation for an intricate problem has taken almost two centuries of scientific research, which also taught us that time is a key component of evolution. The quest is now quite substantially solved, with the group of scientists gradually but constantly working on it. The framework for the discovery applies to all the living creatures. Humans are no exception.

The evolution of humans to reach the present situation has taken over 4 million years. It is intriguing when humans originated and from whom. When did they appear to occupy the planet earth? Where did they come from? This apparently happened in some way from evolution in the prehistoric period or even very deep in the past. Of some phases of the appearance of life on earth followed by its complete or partial elimination, the present duration of life has witnessed the appearance and development of human species which passed through three prominent eras.

We have seen that approximately 1600 million years ago the planet earth witnessed the birth of unicellular complex organisms. The following 100 million years' period saw the birth of organisms with a nucleus. They were the first organisms to carry DNA and RNA. During the next 300 million years, due to sexual reproduction, their number increased enormously, at a very fast rate. They had proficiently adapted themselves to the earth's environment. About 1000 million years ago, multicellular organisms developed from these unicellular creatures. With the conspicuous increase in the mass of the earth and development in the atmosphere, the earth's resolution became slow, making the day of 18 hours. This resulted in a noticeable drop in the earth's temperature. The period approximately 600 million years ago, saw the inception of the Ice Age and the following several million years, the entire planet remained covered under a sheet of ice. About 580 million years ago *jellyfish* came into being and about 57 million years ago *Arthropods* were born. By this period there was an enormous amount of oxygen available on the earth. Since most of it was unused, the oxygen broke down to a pale blue gas, named *Trioxygen* or *ozone*, $O_3$, which is much less stable than its parent *diatomic allotrope*, $O_2$. It was unstable in the lower atmosphere, less in concentration and broke down to oxygen. It became concentrated more in the *Stratosphere* and formed the *ozone layer*, which prevented the sun's ultraviolet radiations from reaching the earth surface. This facilitated the growth of life more commendably on the earth. The Arthropods who were the ancestors of present-day insects took birth in the period about 57 million years ago. The following period of a few million years witnessed the birth of fungi and fishes. Till then about 80 to 90% of life had developed in the oceans only. Approximately 440 million years ago about 48% of the organisms vanished from the earth. Almost about 430 million years ago some of the ocean-dwelling creatures had learnt to live on the land also. By this time vegetation had also grown on the land with a variety of trees and plants, inhabiting the land. About 410 million years ago fishes had developed denture and jaw and in forests, land-dwelling creatures like spider, *centipede*, etc. belonging to class *Chilopoda* had taken birth. About 370 million years ago, the majority of the sea creatures got destructed leaving behind just about 30% of the original population. This was followed by the birth of amphibians, who were the ancestors of creatures like frogs, while the oceans were dominated by sharks. Newly-entered ecological niches brought about abundance of species by a process called Adaptive Radiation. Almost at the same time on land, in the forest areas the plant kingdom got new members—spermatophyte, the *Gymnosperms* and *Angiosperms*. They were vascular and woody plants. Again another event of collective annihilation started around 250 million years ago. This resulted in the destruction of about 95% of the sea-dwelling creatures and 30% of land-dwellers. This continued for around 25 to 30 million years. This was followed by the fascinating entry of many species of dinosaurs about 230 to 240 million years ago. These giants ruled the earth for over 165 to 175 million years' period that followed. Approximately 150 million years ago the avian and mammal fauna came into existence and undertook the journey of development. Simultaneously approximately 130 million years ago, flower and fruit-bearing plants and trees had developed, attracting insects and moths, which relentlessly facilitated the process of pollination. The reptiles like crocodiles and tortoise came to exist approximately 110 million years ago. Pollination in plants received a quantum jump by the birth of honey bees and many other insects approximately 100 million years ago. Then approximately 80 million years ago the reptilian and insecta classes got enriched by the birth of snakes and ants. Approximately, 168 million years ago Dinosaur *Tyrannosaurus Rex* had developed and roamed on the earth surface for the following 20 million years. Approximately 65 million years ago a catastrophic asteroid 6-16 km. in diameter travelling at a speed ten times that of a rifle bullet collided with the earth off the coast of México in the north of Yucatan peninsula, where a crater as large as 175 km. is deciphered even on this day. The collision triggered volcanic eruptions and a powerful earthquake. On the outer ring of the crater so formed, the cracks developed in the limestone terrain running deep. This allowed the groundwater to flow through, eroding the rocks and forming the caves. As the limestone rock profusely reacts with the cavernous water, gradually the ceiling of the caves became too thin to resist and collapsed thus creating 'cenote'. The word cenote owes its origin to the Mayan world, meaning the 'sacred well and an entrance to the cave'. Cenote contains super-clean, crystal-clear water and there are innumerable numbers present on the peninsula. Ahead we will see how significant they were to Mayans. This impact had produced enormous quantities of dust and debris as clouds and amply cooled the earth. This also caused the phenomenal extinction of life and was a very significant event in the evolution of earth and the life on it. This is the K-T or K-pg mass extinction, detailed out elsewhere, which was responsible for the elimination of about 80-90% of oceanic and 85% of terrestrial life from the earth. This was presumably the incidence which eliminated the dinosaurs also whose species had roamed the earth for over 160 to 165 million years. Based on the fossils found it may be stated that much larger number and species of dinosaurs lived on deserts and badland of North America, China, Argentina and in smaller number in other places. T. Rex, the largest and probably the most ferocious of them occupied the heavily forested river valleys in North America before its complete extermination due to K-T mass extinction. The earth was now profusely inhabited by mammals. Approximately 40 million years ago whale fishes, the bats and butterflies took birth. During the following 10 million years, wild animals like a variety of cats, deer, giraffes, bears, etc. took birth. Then approximately 14 million years ago the great Apes, also called *Vanmaush,* came into existence on the earth. Large grasslands developed during the period of following 4 million years which offered favourable habitation to some particular life proficiency. Till the period of approximately 5 million years, creatures like Tree Sloths, Zebra, Elephants, Dogs, Rats and other carnivores had developed. The narrative beginning with *primates* spans over the last about 50 million years. It was around that time when the *primates* got bifurcated into two suborders, viz., the *Prosimians* and Anthropoids. The *Promisians*, often referred to as Pre-monkey are principally the Lemurs, members of *Lemuridae* family inhabiting the thick forests of the island of Madagascar in East Africa. There are about 53 species of Lemurs flocking the island even today and are comprehensively adapted to the environment. To some of the biologists, Lemurs resemble the oldest ancestors of *primates* who lived way back much before the mark of 50 million years. Some of the prominent Lemurs known to thrive in Madagascar include Ring-Tailed Lemur, Coquerel's Sifaka Lemur, Diademed Sifaka Lemur, Verreaux's Sifaka Lemur, Red-fronted Brown Lemur, Indri Lemur, Black and White Ruffed Lemur, Common Brown Lemur and Milne Edward's Lemur. The Red-Shanked Douc is the most colourful of the *Prosimians*; mainly for this reason the animal is nicknamed the "queen of primates" also. During the period from 55 to 34 million years ago in Eocene Epoch, monkey-like *primates* witnessed a major evolutionary change that was taking place in their heads. The hole known as *Foramen Magnum* which allows the spinal cord to connect with the brain began to shift from the back of the skull towards the centre. This implies that some of these early *primates* were moving towards walking in a more upright position. Higher *primates* appeared in Oligocene epoch during the period between 34 to 23 million years ago.

The *primates* had started evolving from African Siamins about 50 million years ago. Taxonomically the *primates* are bifurcated into two groups, the suborder Prosimii and Anthropoidea. The members of suborder Prosimii are popularly called *Prosimians* and the ones classified under *Anthropoidea* as Anthropoids.

The apes are classified under superfamily *Homonoidei*. Tracing its lineage through great apes, gibbons, new world monkeys, chimpanzees and bonobos, the humans began their rugged path of evolution. It is this Anthropoidea suborder which we orient on to find the evolutionary lineage of humans. The humans are classified under tribe *Hominini*, who diverged from the lineage of chimpanzees and bonobos classified under separate tribe *Panini*. This divergence took place more than 4 to 5 million years ago when the humans in one line and chimps and bonobos in the other line evolved from a common ancestor. The Old World monkeys genera included baboons and macaques. Old World monkeys diverged from apes about 27 to 30 million years ago. As assessed by molecular estimates, the gibbons, the apes, a cousin of humans, belonging to superfamily Homonoidei had diverged from the common ancestor of humans and apes about 17 to 18 million years ago. Gibbons share 96% of the genome with humans. Gibbons act as the bridge between the baboons and macaques. About 15 to 19 million years ago the ancestors of subfamily *Ponginae* comprising three species of Orang utans split from the main ape line. Considered to be the most intelligent of the great apes, the Orang utans are also taken as the step next to gibbons en-route to humans and are followed by gorillas. Sharing the membership in the *Hominidae* family with chimpanzee and humans, gorillas have about 98% of their DNAs resembling those of humans. Chimpanzees and humans are the closest relatives of gorillas, all of them having diverged about 7 to 10 million years ago from a common ancestor. Human gene sequence differs only 1.6% on average from the sequences of corresponding Gorilla genes, further differing, however, in the number of copies each gene has. Chimpanzees, our ape cousins are the next on the evolution journey. The evolutionary divergence between Chimpanzees and Humans began to occur between 5 to 7 million years ago.

Chimpanzees and humans share 98.7% of DNA but differ in appearance, behaviour and ability. Size of their brain is almost as big as many of the extinct members of the human lineage. The next in the evolutionary line is bonobos also historically called Pygmy Chimpanzee and less often the dwarf or gracile. The chimps and bonobos retained their ancestral traits until they radically split into two separate groups about 1 million years ago. Chimpanzees and bonobos share about 98.7% of their DNA makeup with humans and are thought to be the closest relative of the humans. Unique to Congo, the bonobo apes are thought to be more intelligent than chimpanzees and share 99.6% of DNAs among them, as revealed by one study. The genetic relationship between chimpanzees and bonobos show striking parallels to the evolutionary history of modern humans. Bonobos are considered to have separated from their common ancestors about 2 to 3 million years ago. The first in the true human lineage is thought to have evolved from chimpanzees and bonobos. With certain similarities, the difference is seen in their social conduct. The chimpanzees live in a male-dominated world while bonobos adopt a female-dominated society. Both use stone tools and sharpened sticks. The chimpanzees live in nests while bonobos spread leaves and branches to make their bed.

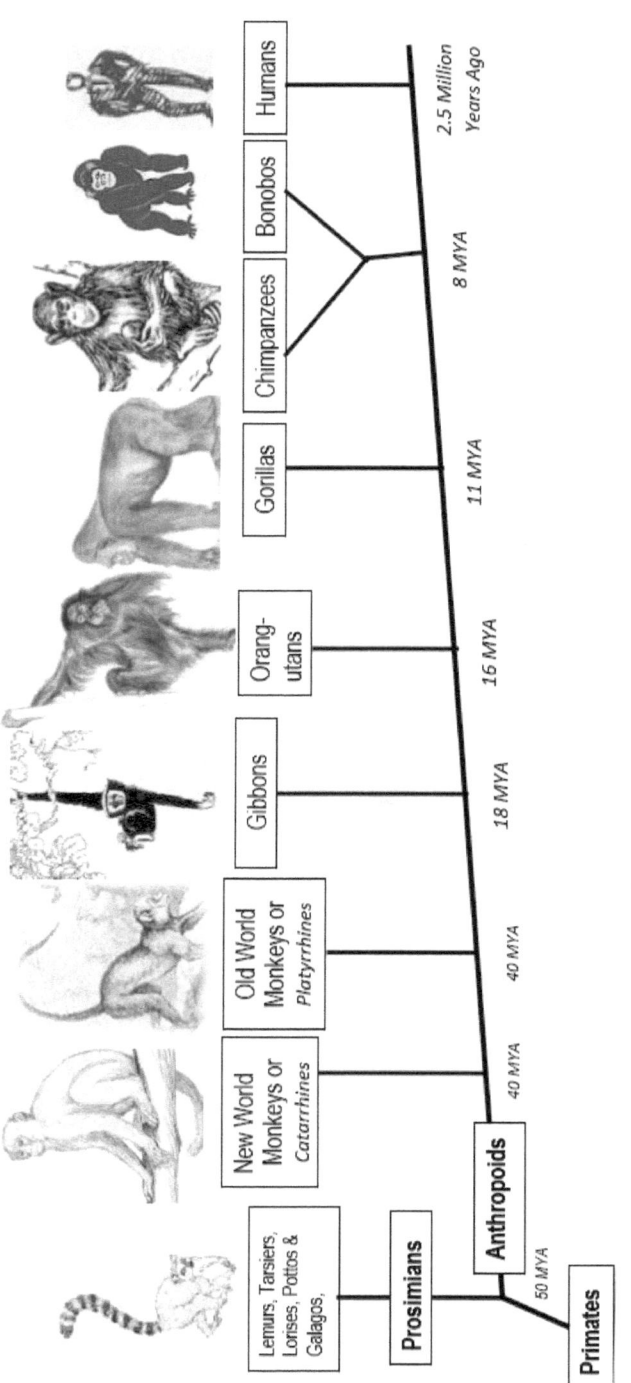

Evolution of Humans; Cladogram of Primates

Around 10 to 12 million years ago, the *primates* split into two major groups from one common ancestor. These two lineages evolved separately to become different species present even today. As we have implicitly seen, the earliest version of members of the group known today as great apes, like gorillas, chimpanzees and bonobos in the African continent are characterised by their arboreal living and quadruped nature. The other group evolved in a completely different way, choosing to become terrestrial and adopted bipedalism. Their brain also increased in size. Later on, through evolution, species named *Australopithecus* and *Ardipithecus* were to start the lineage for humans, passing through different stages.

The study of conspicuous traits of different living apes is taken as the main criteria for taxonomic classification of the group. Many taxonomic classifications of superfamily Homonoidei, which human belongs to, have been proposed by different schools of workers from time to time. The one most accepted widely and the one which represents this group of Anthropoids most comprehensively is graphically produced here. Homonoidei superfamily comprises two families, viz., *Hominidae* and *Hylobatidae* which are further differentiated into sub-families *Ponginae* and *Homininae*. Great apes such as Orangutans belong to the genus *Pongo* of *Ponginae* subfamily. Parallel to this family runs another, Homininae which breaks into two tribes *Hominini* and *Gorilini*. Genus Gorilla is a member of the later, while Hominini includes chimpanzees and bonobos. Almost 99% of human DNAs are shared by the members of Hominini. Biologically, thus they are the closest relatives of the humans. The humans who evolved have not reached the present form in a short span. Tracing the lineage of present humans, belonging to the genus Homo, is also very intriguing. Once again the fossils record maintained by nature is the only source. The term *"Hominid"* is often used to describe all the bipedal, erect walking *primates* which may or may not be necessarily human. The shift from apes to humans was sluggish which is taken as a logical explanation for evolution.

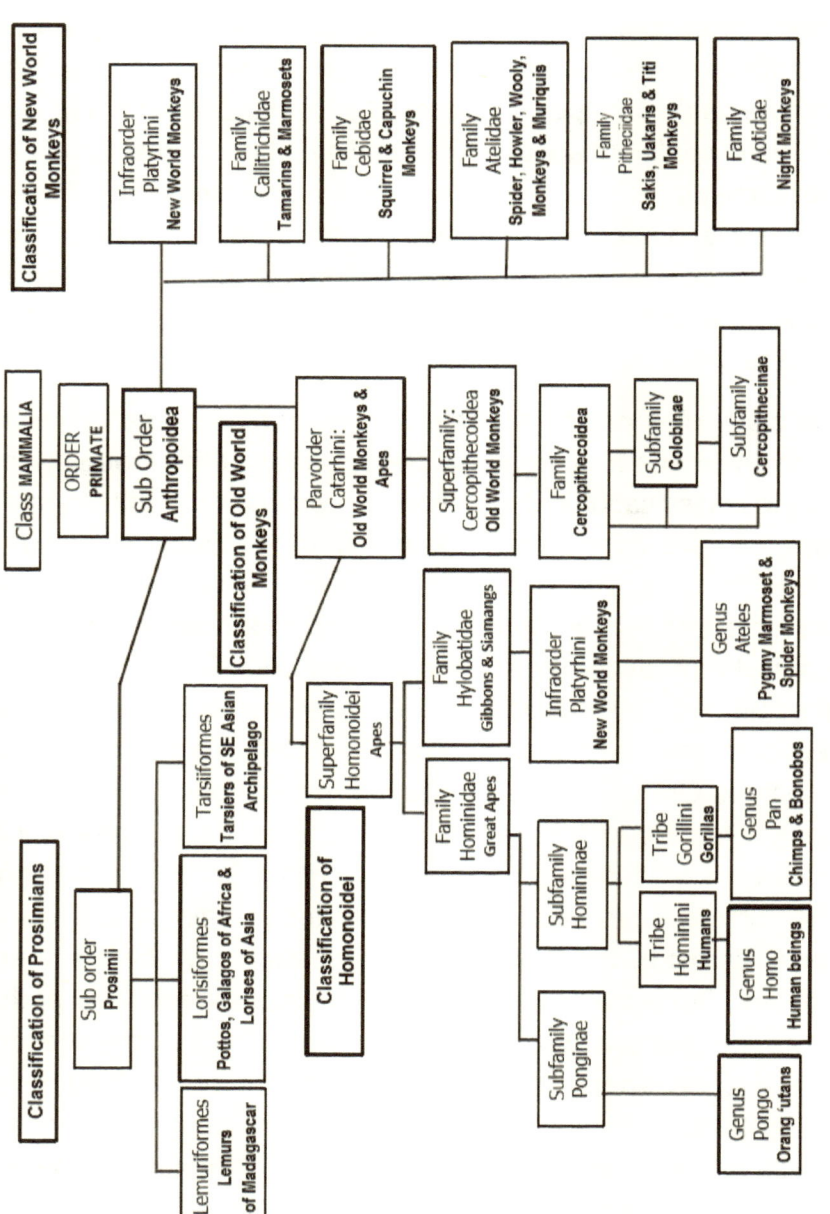

Taxonomic Classification of Order Primate

It is now established that about 99% of the DNAs of chimpanzees and humans are common and so are the traits. Chimps, as they are often called, resemble some other species in the family tree of humans close to the roots down below.

Now that the first step towards evolution to become the human as you and I from chimpanzee is taken and the occupant of the lowermost part of the human family tree has arrived, the lineage has begun. It is the *Ardipithecus group* from where we start climbing the family tree. It is the genus of an extinct hominine, that roamed the planet during the period from early Miocene to early Pliocene. A rather complete skeleton of *Ardipithecus ramidus*, nicknamed "Ardi", a female weighing about 50 kg., with a small brain, was retrieved from Afar desert of Ethiopia at a site known as Aramis in middle Awash River valley region of Ethiopia, complete to the tune of about 45%. The radioactive dating of the volcanic ash encasing and hosting Ardi's fossils suggested that she lived about 4.3 to 4.5 million years ago. This species acquired its name from Ardi of Afar language meaning 'floor or ground' while pithecus stands for "ape" in Greek. It was initially thought to be one of the earliest ancestors of humans, after they diverged from the members of genus Pan, mainly the chimpanzees. At one point of time, the matter of whether they qualify to be Hominini at all became fiercely debated. Two fossil species *Ardipithecus ramidus,* and *Ardipithecus kadabba* represent the genus. Initially, the behavioural analysis showed that they could be very similar to chimpanzees which was later proved to be otherwise. By basic expanse of analysis, it can be stated that the early ancestors of humans were very chimpanzee-like in behaviour. *Ardipithecus ramidus* differs from chimpanzees as the latter's feet are specialised for grasping tree branches while *Ardipithecus ramidus'* feet, on the contrary, are better suited for walking. This species was bipedal and probably quadrupedal on trees, an interpretation supported by its 'big toe' or grasping hallux, adopted for locomotion on trees. It had reduced canine teeth and possessed a small brain measuring about 300 to 350 cm$^3$. It was more Prognathic than modern humans. This species was omnivore and frugivore. The canine in both male and female were of the same size and were not suited for hard and abrasive food, a specialisation adopted by other apes. *Ardipithecus kadabba,* named from the Afar word meaning "basal family ancestor", is described as "probable chronospecies of *Ardipithecus ramidus,"* and was originally thought to be its subspecies. On the basis of the discovery of teeth from Ethiopia, it is now thought to be a species distinct from *Ardipithecus ramidus.*

*Sahelanthropus tchadensis* also clubbed with the *Ardipithecus* group, chronologically occupies the place closest to the roots in the family tree, is an extinct species of *Homininae.* Thought to have existed during Miocene epoch, disputably about 6 million years ago, *Toumai* is the nickname given. The fossil which pronounced its existence was discovered somewhere in the desert zone in the north, close to the Sahelian arid zone in the Republic of Chad. It is thought to have lived close to the time when genus human diverged from genus Pan—the chimpanzees and bonobos. Highly controversial but true sensation unearthed in Tugen Hills region, a part of Africa's Great Rift valley, geographically located in central Kenya, when French Geologist Martin Pickford and Palaeontologist Brigitte Senut discovered about 11 human fossils on the floor bed of an ancient water body. As before, once again, Africa lived up to its epithet of the "Cradle of Humanity". These fossils dated between about 6.2 million and 6 million years in the past. The luck had in fact, favoured renowned fossil hunter Kiptalam Cheboi as it was he who had retrieved the jawbones of a male later to be named *Orrorin*. He was also given the nickname of Millennium Man. *Orrorin* was definitely closer to human traits than those of apes. With its age surpassing the oldest fossil of human-like creature found at that time, it had all reasons to be an astonishing discovery. *Orrorin tugenensis* was a bipedal but climbed the trees possibly with similar ease.

In order that the age of the fossilised male skeleton is established, the local geologists studied the area closely. The area known as Lokeno formation is an ancient lake basin. Volcanism in two separate phases over an area of about 50 x 60 km. all around resulted in the older flow of *Trachyte* followed by *basaltic* lava flow. *Orrorin* was probably sandwiched between these two flows, which on disintegration and weathering of these igneous rocks, was set free and came to be deposited in the lake basin.

This finding attracted controversies as it was going to change the basic thought of human evolution history. There were many grounds for apprehension and the onus to prove their thought was on the working school of scientists. The only solution was if the age of *Orrorin* could be established by a reliable method. Earlier discoveries in Kenya had proved much younger than claimed. Another was that this fossil may have been carried by surface water from much younger formation of rocks and deposited where it was recovered from, a dried watercourse. The supporting feature also came along; *Orrorin* rested in the company of elephants, horses, hippos, etc., whose fossils were definitely clearer. It, therefore, became incredibly clear with a more powerful judgement that the assemblage was that of Miocene pegged at about 6 million years. Ultimately Shimane Medical University, Matsue of Shimane Prefecture in Japanese archipelago of Honshu, Japan gave out an unchallenged verdict that *Orrorin* comes from a period of 5.8 to 6.1 million years. They used the earth's magnetism and the tiny magnetic particles in the rock that enclosed the *Orrorin* fossils for arriving at the conclusion with authority. The oldest Hominid before *Orrorin* was about 4.2 million years old which meant that *Orrorin* was now the oldest.

The process of evolution of humans, according to one school of scientists, began about 6 to 7 million years ago. *Ardipithecus* group fossils ranged from about three million years to about 6-7 million years back from the present. The range as proposed incorporates the fact that *Sahelanthropus tchadensis* was considered, an early

Hominid by one school, while the other school propounded that it was just an ancient ape. The fact, however, cannot be denied that it adopted bipedalism and had small canines. This discovery led to an inevitable inference that human lineage had started diverging, parting the company of the lineage of apes. Nevertheless, this period had an impeccable store of scientific information on the evolution of genus Homo which acted as the springboard to clip the greater heights. As we move upwards the next group we come across is *Australopithecus* group.

This group included four species, namely *Australopithecus anamensis, Australopithecus afarensis, Australopithecus africansis* and *Australopithecus garhi. Australopithecus anamensis* is a group of hominin species that existed between 4.2 and 3.8 million years ago. The first fossils were stumbled upon in Kenya in the area near Lake Rudolf, now known as Lake Turkana in Kenya and Ethiopia. This was regarded as the oldest hominin found on the planet before the claim was successfully staked by *Orrorin Tugenensis*. However, *Australopithecus anamensis* is the oldest known among the group. *Australopithecus* afarensis is an extinct hominin and is considered to be the direct ancestor of humans and other *Australopithecus* species. It was slender in build and lived between 3.9 and 2.9 million years ago. The first fossil assemblage of *Australopithecus* afarensis was discovered by Donald Johanson, an American paleoanthropologist in Hadar Triangle region of Ethiopia. Johanson was assisted by a graduate student Tom Gray during this monumental work. This fossil was discovered about 40% complete at the time of its finding. This was considered to be the oldest at that time and was nicknamed 'Lucy'. It was a three-and-half feet female whose anatomy was in a transitional state between apes and humans. The mainstream school of scientists use this as evidence that humans evolved from Australopithecines who in turn evolved from apes. The fossil of *Lucy* was complete almost 40%, whose discovery created a sensation among the scientific fraternity the world over. *Australopithecus* afarensis is an extinct species of the Autralopithecus, who lived aboto 2.1 million years ago. It was slender in build or gracile. It was thought to have been a direct ancestor of humans. The fossils were discovered in Taung, Sterkfontein, Makapansgat and Gladysvale cave in South Africa.

Fossils of a male child named *Taung* was found in a limestone quarry at Taung near Kimberley. Mrs.Plea, later understood to be a male was retrieved from Sterkfontein caves of limestone in Gauteng province near Johannesburg. Sterkfontein is a South African National Heritage site and the area carries an epithet *"Cradle of Humankind"*. Paleoanthropologist Ronal Clark discovered the fossil remains of a male *Australopithecus* named "Little foot" in Sterkfontein area. *Australopithecus* was bipedal hominin as evident from its pelvis, with arms longer than legs to a certain extent, similar to a chimpanzee. It exhibited the traits of both human and apes. Its brain measured about 400—500 $cm^3$. Currently, the Little Foot is classified with *Australopithecus* but the dispute still continues whether it is an independent species or an *Australopithecus*.

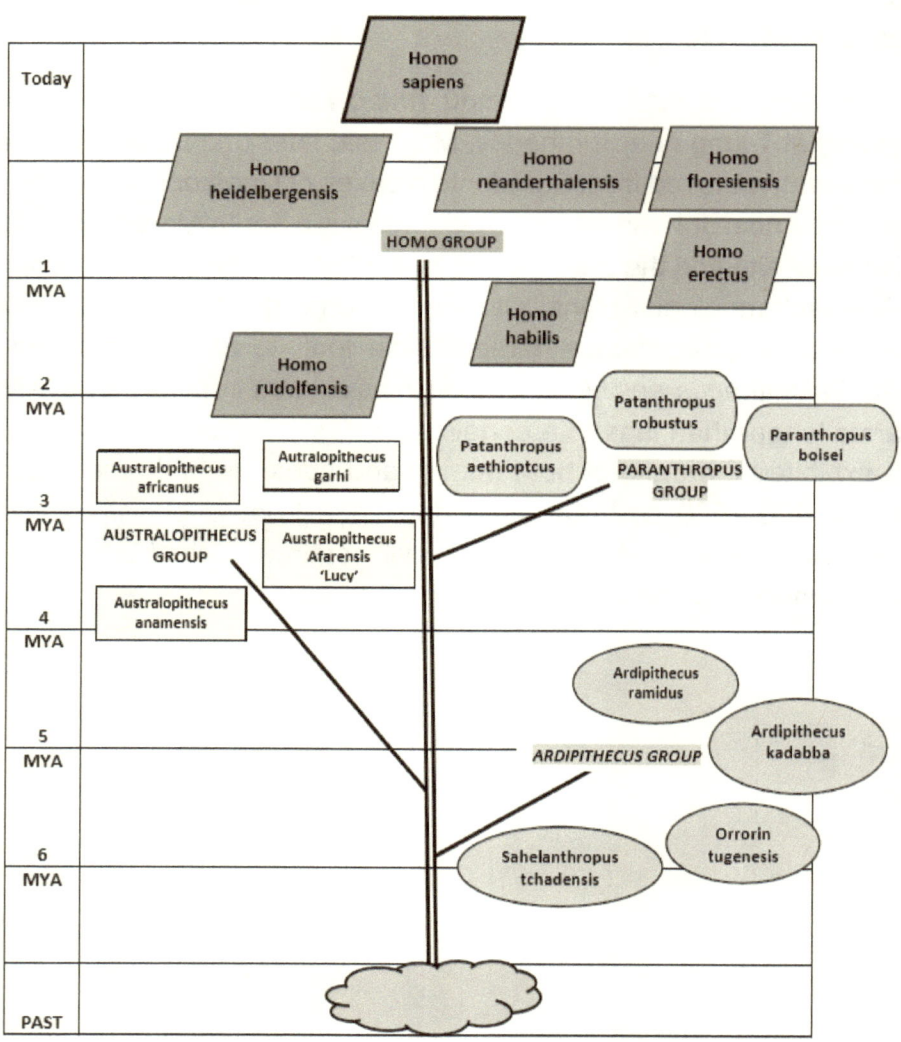

Family Tree of Genus Homo

*Australopithecus garhi* lived about 2.6 to 2.5 million years ago. The fossil remains of *Australopithecus garhi* were first retrieved from Bouri formation in the Afar Region of Ethiopia. Like others, it was also thought to be the direct ancestor to Homo in the human line. Its brain was 450 cm$^3$ and had large molars and pre-molars, had adopted bipedalism and arborealism. They were the first who are postulated to have manufactured tools. They probably produced the Oldowan industry which is a tool-making tradition characterised by crudely chiselled rocks and pebbles and was pioneered during prehistory.

As we scramble up the family tree, moving away from the roots, the group that we come across is the *Paranthropus* group. The word *Paranthropus* meant "beside man". *Paranthropus boisei, Paranthropus robustus* and *Paranthropus aethioptcus* are clubbed in a group named *Paranthropus* group. They occupy a position above *Australopithecus* but below the Homo group. *Paranthropus robustus* is an extinct hominin discovered in South Africa in the late thirties. The cranial features in *Paranthropus robustus* were prominent and had developed for "heavy-chewing". Fossils of *Paranthropus robustus* were excavated from South Africa which comprised over 100 specimens retrieved from limestone cave of Swartkrans. One school considers it an *Autralopithecus* and has named it *Australopithecus* robustus. A well-preserved cranium of *Paranthropus boisei* was discovered by anthropologist Mary Leaky in Olduvai Gorge, Tanzania sometime in 1959. It had large flat cheek teeth with thick enamel, and was appropriately nicknamed '*Nutcracker Man'*. In addition to Tanzania, the fossils of *Paranthropus boisei* have been discovered in Ethiopia and Kenya also in East Africa. It was initially named *Zinjanthropus boisei*, after the East African region of Zanj, and boisei after the benefactor of the exploration team, Charles Watson Boisei. Its age was placed at 1.75 million years ago, estimated by potassium-argon dating of anorthoclase crystals found in the volcanic ash, the tuff bed which overlay the stratigraphic horizon which encased the fossil skull. An adjoining volcano poured the pyroclastic ash flow which might have buried the skull after killing the Nutcracker. The size of their brain was not big either, ranging from 500 to 550 cm$^3$. This species became extremely important after Richard Leaky of legendary Leaky family and the son of Louis and Mary Leaky, highly acclaimed anthropologists, suggested that it was the first hominin to use the stone tools. The stone tools recovered from Ethiopia were dated to 2.5 million years ago in Liverpool University laboratories and were possibly made by *Paranthropus boisei*. Later a complete jaw, named as Penini mandible was discovered by Kamoya Kimeu, regarded as world's most successful fossil hunter, in the decade of the sixties in Penini, Tanzania. Richard also discovered another skull at Koobi Fora near Lake Turkana in Kenya, thought to be the nursery of ancient apes and human fossils. *Paranthropus boisei* became equally ill-famed for transmitting *genital herpes* to the species Homo which continues till this day. This came from the research conducted on viruses by Cambridge University and Oxford Brooks University and propounded that this heavy-set, bipedal with a rather small brain was the culprit. This happened between 3 and 1.4 Million years ago. In fact, humanity had ducked the virus but HSV2 jumped the species barrier from African Apes back to human species through an intermediate hominin species not related to humans. *Paranthropus aethiopicus*, was named after its first discovery in Ethiopia. Later its fossils were discovered in Kenya also. Omo 18 the first fossil find was discovered in Ethiopia by French anthropologists Camille Arambourg and Yves Coppens in the late sixties. This was followed by the discovery of the most famous fossil of *Paranthropus* aethiopicus by Alan Walker. This fossil is also known as "Black Skull" or the *KNM-WT 17000* and was named after its finder as *Paranthropus* walkeri. Fossils from Kenya and Ethiopia were dated to approximately 2.5 million years before present, a time scale used to specify when the events took place before the advent of practical radiocarbon dating in the decade of fifties.

As we go up the family tree and reach the top, a place occupied by group Homo is encountered. It comprises seven important species, viz., *Homo rudolfensis, Homo habilis, Homo erectus, Homo heidelbergensis, Homo neanderthalensis, Homo floresiensis* and *Homo sapiens.*

*Homo rudolfensis* is an extinct species of *Hominini tribe* occupying the bottom of the Homo group. It marks the morphological boundary between Homo and Australopithecus. The fossils thought as those of Homo rudolfensis are rather small in number and are largely inconspicuous. The first fossil of tribe Homo is dated to about 2.8 million years ago. Bernard Ngeneo a constituent member of a team led by Anthropologist Richard Leaky and Zoologist Meave Leaky, while exploring Koobi on the east side of Lake Turkana in Kenya, the first fossil of *Homo rudolfensis* was discovered. It was lectotype and was dated to about 1.9 million years old. It is a matter of debate which continues even at present whether this species should be classified within the genus Homo or *Australopithecus* or a divergent subspecies of initial *Homo erectus.* These fossil remains are named *KNM-ER 1470,* 2012 Koobi Fora finds, *UR 501* and *LD 350-1*.

The fossil *KNM-ER 1470* attracted a lot of attention, particularly as to which species it should be classified with. Its discoverer Timothy Bromage put the cranial capacity at 700 cm$^3$. Some thought it to be *Homo habilis*, and considered it to be nearly 3 million years old. Later it was placed at 1.9 million years as the skull fossil was found to be profoundly different from other *Homo habilis* specimens. This led to the presumption that it was *Homo rudolfensis*. Nothing, however, can be stated if *Homo rudolfensis, Homo habilis* or any other lying murkily somewhere was ancestral to later Homo species.

Meave Leaky discovered three more fossils in Northern Kenya, which included two jawbones and a face. The face, named *KNM-*

*ER-62000* was of a juvenile and matched with *KNM-ER-1470,* which prompted the interpretation that it was a separate species. The age was worked out to 2 million years ago. It thus becomes contemporary with *Homo habilis* and led to a firm interpretation that these two, *Homo rudolfensis* and *Homo habilis,* together with *Homo erectus* roamed about in eastern Africa in the early *Pleistocene epoch.* UR 501, a jaw bone was discovered in Urah, Malawi and was taken to be that of Homo habilis, dated to 2.4 million years ago. LD 350-1 was discovered in Ethiopia, dated to about 2.8 million years ago and thought to be closely related to *Homo rudolfensis.*

A fossilised mandible almost complete with all appengdages, discovered from a sand quarry in Rosch, near Heidelberg in Germany. German scientist Otto Karl Friedrich Schoetensack described it as a heavily built individual, with a large brow bridge, a large braincase but lacked a chin. It was named Homo heidelbergenasis whose body structure could be described as wide and short. They were further ascribed to have acquired full knowledge of fire and its control. They used wooden spears and pioneered large animal hunting. The size of their brain was almost the size of that of humans, ranging from 1206 to 1230 cm$^3$. Another striking feature in their body construction was that they completely lacked the airsacs, the *laryngeal diverticula* which might have been the probable cause for development of the vocal language which presumably they had. It was the first species to build shelters and had learnt to make dwellings using rocks and wood. Shunning the tendency of 'nomadism', they spread across countries like Ethiopia, Namibia, South Africa, England, France, Germany, Hungary, Spain, etc. They had developed the practice of burying their dead with respect, probably the oldest such practice in the family tree of humans. One school of scientists believe that modern humans are derived from Homo heidelbergensis via an obscurely described species Homo rhodesiensis. However, on the basis of genetic analysis of fossils discovered in Spain, it was more prudent to consider the Homo heidelbergensis in the lineage of *Neanderthals*. The members of this species whose fossils have been discovered from China were named Dali Man and Maba Man.

Homo habilis, also nicknamed as 'handyman', is one of earliest members of genus Homo. The species was discovered by anthropologists Louis Leaky and Mary Leaky in Olduvai Gorge area in Tanzania associated with the Oldowan stone tool working in the decade of fifties. This species inhabited south and East Africa. The South African Raymond Dart christened the species by the name 'habilis', which in Latin stands for someone 'handy, able, mentally skilful, vigorous'. Its braincase was slightly larger but the face and teeth were smaller compared to *Autralopithecis*, the older hominin species. *Homo habilis* is an archaic species that lived between about 2.3 to about 1.65 million years ago during the early period of *Pleistocene*. It was thought to occupy a position between *Austrlopithecus afarensis* and *Homo erectus.* It hunted more frightening carnivores. Later on more fossils were discovered but they turned out to attract more of debates rather than a support for already conceptualised ideas.

*Homo erectus* took over the stage of evolution following the *Homo habilis.* Nicknamed "upright man" *Homo erectus* is an extinct species that walked erect and was bipedal. This species inhabited Northern, Eastern and Southern Africa, Dmanisi, the Republic of Georgia in Western Asia, Indonesia and China in eastern Asia. In fact, it is considered to be the first species to have expanded beyond the African continent. It was a highly variable species that was possibly the longest living early human species, much longer than *Homo sapiens*, our own species—roughly about its nine times. Eugene Dubois, a Dutch surgeon discovered the first fossil in Indonesia in 1891. Around the same time fossils were found in Java and the same was named "Java man". Then "Peking Man" was discovered in China in 1920. They comprise classic examples of this species. However, the fossil assemblage discovered in Dmanisi of Georgia is a true museum of *Homo erectus* fossils. The most complete fossil of the species, named 'Turcana Boy' was found in Turkana basin area in Kenya. It is dated to about 1.6 million years ago. This species lasted between 1.89 million to 110,000 years ago, that is spanning through most of the Pleistocene epoch. Regardless of Dmanisi fossil's precise identity, researchers agree unanimously to the age. It is amply clear that Homo habilis and *Homo erectus* probably roamed together on the planet, maybe for several of their generations. The wealth of fossils and artefacts discovered offer rare evidence for a critical moment in the saga of human evolution. Mysteries remain, yet strong lessons emanated. Georgian Anthropologist David Otarisdze Lordkipanidze appears to be grateful as he is quoted as saying, "I want to thank the people who died here."

*Homo erectus* had a brain of the size between 550 cm$^3$ and 687 cm$^3$, larger than that of *Homo habilis*, was taller, slender and strongly built. Based on the analysis of cranial base analysis of their skull fossils it is clearly interpreted that they could speak. They used more sophisticated *acheulean tools* such as hand axe than earlier species. Use of more effective tools, certain other traits together with higher intelligence and better adoptive mechanisms promoted *Homo erectus* to adopt hunting and gathering. Their diet included meat, nuts, fruits and berries. They had also learnt to make and control fire. Since *Homo habilis* and *Homo erectus* coexisted, an isolated subpopulation of *Homo habilis* may have evolved into *Homo erectus*, and the rest of the subgroups remained unchanged as *Homo habilis,* until their extinction.

The fossil of what came to be known as *Homo floresiensis*, nicknamed "Hobbit" was discovered in Liang Bua cave on the island of Flores in Indonesia. An archaic human species, it is small and inhabited Flore island until the advent of modern humans about 50,000 years ago. The height of the individual would have been about 101 cm. Skeletons of nine individuals were recovered with one complete skull. This species became a subject matter of intense research by the competent fraternity to determine whether they were distinct from modern humans. A well-founded conclusion that they were a distinct species is now drawn on the basis of genetic and anatomical traits. The age bracket of their existence was also worked and reworked several times. Of them, the last and probably the most reliable one places them from about 60,000 to 1000,000 years ago. This period matches quite reasonably with the archaeological horizons ranging from about 50,000 to 190,000 years ago, from which the stone tools were recovered alongside the skeletal remains of hobbits. *Homo floresiensis* also figures in the list of some of the mysterious extinct members of the genus which is produced in the text ahead.

As already pointed out, the period from 3000 BCE to the present is classified as history which is further divided into Ancient age from 3000 BCE to 476 CE, Medival Age from 476 CE to 1492 CE, that is about 5th to 15th century, Modern Age from 1492 CE to 1789 CE while the contemporary Age is represented by the period from 1789 CE to 2019 CE. This entire period virtually witnessed the evolution of various species of tribe *Hominini* of family *Hominidae.* Thus as we inch towards the modern humans, moving up the family tree, we come across another extremely important individual, the *Neanderthals*, member of species *Homo neanderthalensis*. These hominins like other human species originated in Africa and migrated to Eurasia before other humans did. They inhabited Europe, South-western Asia and Central Asia. As many as 500 fossils of *Neanderthals* have been discovered the world over, some unveiling fine to finer details leaving some mysteries behind, unresolved so far. It was named by Geologist William King after these fossils were found in Feldhofer cave of Neander valley in Germany; the suffix 'tal' is the rectified form of 'thal' meaning "valley" in German. The first specimen named Neanderthal 1 was discovered in 1856 in Germany. Later on, more Neanderthal fossils were discovered. Nothing like these was discovered by the scientists before. Though the fossils discovered from Engis in Belgium and Forbes in Gibraltar were the first corporeal remains of *Neanderthals* encountered, they could not be classified taxonomically. This became possible only after the one discovered in 1856 in Germany. After that over 500 specimens of Neanderthal fossils have been unearthed. They were characterised by prominent brow ridge above their eyes, distinctive face, wide big nose and the face which protruded forward. They were hunter-gatherers and did not adopt any domestication and agriculture. A short-time debate cropped up among the anthropologists that the *Neanderthals* were a distinct species of *Homo sapiens* or a subspecies thereof. Our well-known closest relatives fossilised for long now, roamed the earth about 200,000 to 30,000 years ago sharing it with us for some time before getting extinct. It was the Pleistocene epoch, the Ice Age, which probably forced them to take shelters in caves from harsh weather in Eurasia leaving the trail in the form of fossils in the caves. This gave them another nickname, the "cavemen". Their body structure seems to have adapted for harsh cold weather. Their short, stockily built bodies and a wide nose are some features that are quoted in favour of their successful adoption. Their flaring, funnel-shaped chest, flaring pelvis, the robust fingers and toes established them to be different from other Homo species. The brain of *Neanderthals* was almost of the size of *Homo sapiens* today, even bigger sometimes.

They lived in nuclear families, took care of their sick and the ones who were not able to care for themselves. The age normally was about 30 years though some lived longer. Although their large and complex brain suggests that they might have had a language, yet it is neither confirmed nor is proved otherwise. They frequently used stone tools such as scrapers, blades, etc. like other early humans did before. In due course of time, their later generations learnt to make complex tools utilising bones and antlers. According to Evan

Hadingham the scientific editor of *PBS of USA,* the *Neanderthals* used some kind of glue to stick stone chips to wooden shafts for making spears for large game hunting. *Neanderthals* knew about controlling the fire and are said to adopt cooking their food. *Neanderthals* were primarily carnivorous. It is even theorised they built boats and sailed on the Mediterranean though the motive behind it remains unexplained by the proposers of such ideas. They also adopted a sophisticated way of life and respected their dead by burying them. Out of over 500-odd discoveries of *Neanderthals* the world over, only 30 to 40 sites have good burial evidences. Some of the Neanderthal burial sites include La Chapelle-aux-Saints, Central France and a huge complex mass burial site in La Ferrassie in Dordogne region of France. Shanidar limestone cave in Iraq which houses cemetery with as many as about 35 individuals of different age groups presents a more convincing picture of the burial practices adopted by *Neanderthals*. Even though such ratio probably cannot be taken as their being habituated for adopting such practices, there are other sites out of those 500 ones that squeak a completely different story. In fact, the *Neanderthals* resorted to *cannibalism* also. The sites such as El Sidron, Spain; Gran Dolina, Central Spain; Zafarrya, province of Grande on the Iberian Peninsula; Moula Guercy, Los Angele county, California; La Pradelles, France; and Goyet, Belgium are prominent. It has not been possible to work out till date whether such gruesome practice was a part of rituals or it came to satisfy the hunger.

The subject matter whether or not the *Neanderthals* interbred with other human species is fiercely debated. In fact, 99.7% of *Neanderthal DNA* is identical to that of modern humans. It is also found that 2.5% of an average non-African human's genome is made up of *Neanderthal DNA* whereas the average modern African has no *Neanderthal DNA*. This set of information supports the interbreeding hypothesis; it suggests that they bred only after the other humans had moved out of Africa, into Eurasia which means this could have happened as recently as 37000 years ago. Another school of scientists, questions the interbreeding hypothesis and rules out that two species ever lived on the earth together. The Neanderthal bones from southern Spain were 50,000 years old while modern humans are not believed to have settled in the area until 42,000 years ago which meant that it is unlikely that they were ever together. If this is so, how did the similar genomes come into being? This probably could have been the result of both the *Neanderthals* and modern humans evolving from a common African ancestor, a hypothesis that looks more prudent and allows no flight of imagination. The *Neanderthals* went extinct in spite of cautiously and naturally crafted lifestyle. Until recently the idea that they interbred with modern human species until they were absorbed into the latter's, was popularly accepted. Some thought that modern humans, being more ambitious killed the *Neanderthals*. Still others believe that it was a spontaneous or dramatic or gradual climate change and their inability to adapt to the changes in the environmental set-up that drove them out to extinction.

We have climbed the human family tree, an upward stride in time which started about 5 to 6 million years ago and are close to the tree canopy. However, some anthropologists have rightly opted for a more studious approach and stumbled upon a short time stopover in their journey from *Neanderthals* to species *Homo sapiens*. This stopover was the *Cro-Magnons*. The *Cro-Magnons*, however, are now thought of as a cultural entity slightly diverged, not as a separate species and are classified with humans in tribe *Hominini*. The first fossil skull named *Cro-Magnon 1* was discovered at the famous rock shelter site at *Cro-Magnon*, near village Les Eyzies in France. This place also had fossils of four adults tucked away in this limestone rock shelter. It was intentional burial, which was endorsed by the presence of shells, animal teeth and stone tools. The tools and other material associated with the site have been dated to *upper Pliestocene*, probably between 30,000 to 32,000 years. *Cro-Magnon 1* might have been about 50 years in age at the time of death. The traits that are interpreted from the skull, almost complete except the teeth, are unique to modern humans. They were quite similar to modern humans anatomically. They shared the earth with *Neanderthals* and probably with modern humans as well, as thought by one school of scientists. The most striking feature was their cranial capacity. Some of the characteristics of Cro-Magnons differed from modern-day humans. They had a robust physique and a larger cranial capacity, fairly low skulls, fat mandibles, blunted chins, well-shaped noses and with no or moderate prognathism. Their eye orbits were significantly rectangular, matching with modern Ainu people and many sub-Saharan Africans. The vocal apparatus was similar to those of modern humans. Their cranial capacity was 1600 $cm^3$, which is much larger than average modern humans. *Cro-Magnons* apparently led a physically tough life. Some adults had fused vertebrae in the necks, a sign of traumatic injury. The adult female had skull fracture before and had apparently survived for some time after that. Their survival after receiving such serious injuries clearly endorses that the group supported such members and cared for them. As a much shorter period has elapsed while they roamed the earth, it is thought easier to decipher the causes for their extinction. They probably got absorbed gradually into the modern human by virtue of interbreeding. There are many other fossils which signify various characteristics after the discovery of *Cro-Magnon 1*. *Cro-Magnons* were frequent users of flint tools which are clearly associated with *Aurignacian culture,* the tool-making industry and tradition of artistry adopted by Upper Palaeolithic Europe, involving fine working on bones or antler points. They used such articles for making body ornaments and flints for making armaments for hafting, serving a more effective purpose. *Cro-Magnons*, virtually *Homo sapiens*, were successful big game hunters due to their improved tools. They were frequent users of fibre. They spun, dyed and knotted flax and made artefacts recovered in Dzudzuana. Bones of mammoths, clay, branches and animal hide were used for making shelters and for fighting the inclement weather conditions. Rock paintings were drawn and painted by using manganese and iron oxide, etc.

As can be possibly made out from the course of preceding portrayal that the study of an intricate subject like evolution must certainly have involved an interdisciplinary approach. The fact is that the scientific disciplines of anthropology, primatology, palaeontology, geology, archaeology, zoology, botany, ethology, linguistics, embryology and genetics cumulatively play their respective roles for enriching the knowledge of evolution. Evidently, all the conclusions require interpretations and anticipations. Thus the element of mistakes, inaccuracies and misconceptions cannot be ruled out, at least not completely. Still, it is brought to notice that the species of *Homo sapiens* have provided the modern scientific fraternity with the largest number of fossils, intact, easily interpretable, representing almost all the stages of development of the species and a span of past about 300,000 years. In the *Pleistocene epoch*, between 400,000 years ago and the second interglacial period around 250,000 years ago, expansion of intra-cranial space and advancement in stone tool-making, paved the way for some earlier species like *Homo erectus* or *Homo heildelbergensis*, to evolve into *Homo sapiens*. *Homo erectus* eventually got replaced both within and out of Africa. The theory elaborating this migration and origin is called the "recent single-origin hypothesis" and sometimes "out of Africa". Then there was likely interbreeding between *Homo sapiens* and *Neanderthal* and some other species in Africa and Eurasia, a highly debatable observation.

*Homo sapiens* evolved about 300,000 years ago in Africa, more during the dramatic climate change. Like other early species, they gathered and hunted food. They were efficient and were makers and users of more sophisticated tools, compared to earlier human species. At the same time, they also developed behaviour that helped them to adjust and win over the vagaries of an unstable environment. It was probably this particular trait which endowed them with better chances of survival over other species. During a time of dramatic climate change 300,000 years ago, the fossilised skull of *Cro-Magnon 1*, discovered in 1868, by the French Geologist Edouard Lartet referred to in the earlier text, is considered to be the first one. No fossil of *Homo sapiens* is considered as being the one that gave them their name. Palaeontologist Robert Bakker opined that the skull discovered by acclaimed American Palaeontologist and Herpetologist Edward Drinker Cope was "lectotype". Generally, modern humans are characterised by the lighter build of their skeletons compared to earlier humans. Modern humans have a larger cranial capacity. The average size of their brain is approximately 1300 cm$^3$ and it varies from population to population and between males and females. The skull also got reorganised into a thin-walled, high-vaulted skull with a flat and near-vertical forehead. Either no or much less of heavy brow ridges and prognathism compared to early human species also featured. The jaws of modern humans are less heavily developed with smaller teeth. Sometimes the term "anatomically modern *Homo sapiens*" is used to denote the members of our species who lived during prehistoric times. We do not know everything about our own species—but we keep on learning more and more. We attain more and more details about who we are through studies of fossils, genetics, behaviour and biology of modern humans which goes on unabated. This has truly been aided by some remarkable and sustained studies of our closest ancestral biological relatives, the chimpanzees conducted by the outstanding, superlative and gritty scientific luminary, Dr. Jane Goodall.

The record of fossils of *Homo sapiens* is quite big. The list is given ahead:

| PROMINENT EARLY *HOMO SAPIENS* FOSSILS AND SITES ||||
|---|---|---|---|
| Sr. # | COUNTRIES | LOCATIONS | TIME OF FOSSILS *(tentative)* |
| 1. | East Africa | Herto, & | 160,000-154,000 |
| | | Middle Awash, | --------Do-------- |
| | | OMO 1 | 195,000 |
| | | Laetoli | 120,000 |
| 2. | South Africa | Border cave | 115,000-90,000 |
| | | Klasies River Mouth | 90,000 |
| 3. | Israel | Skhul & Qafzeh | 92.000-90,000 |
| 4. | Australia Asia | Lake Mongo | 60,000-46,000 |
| 5. | **Asia**-Laos | Annamite Mountains | 63,000 |
| 6. | Mongolia | Ordos | 40,000-20,000 (?) |
| 7. | China | Liuijiang | 139,000-111,000 (?) |
| | | Zhirendong | 100,000 (?) |
| | | Zhoukoudian upper cave | 27.000 |
| 8. | **Europe**-Romania | Pestera cu Oase | 36,000-34,000 |
| 9. | Czech Republic | Mladec and Predmosti | 35,000-25,000 |
| 10. | France | Combe Capelle | 35,000-30,000 |
| | | Cro-Magnon | 27,000-23000 |

The migration of humans began into the lower latitudes of East Asia at least 70,000 years ago, unlikely interbreeding with *Neanderthals,* an apprehension occasionally expressed. One might see the genetic markers from this conjugal alliance in New Guinean, southern Cinese territory and people inhabiting Micronesian island. The dispersion of *Homo sapiens*, evolving in Africa dispersed gradually across the planet in due course of time. Scientists have debated over this aspect a great deal and propounded several theories and suggested various models. Some important of them are mentioned briefly. They are, Replacement Model Argument or genetic diversity; Regional Continuity Model Arguments; and the Assimilation Model.

With the discoveries of *Lucy* and *Orrorin Tugensis* belonging to species *Australopithecus*, our understating of human evolution has been relentlessly pushed back in time. Now it is a calendar that stretches to over 6 million years. In present time with so much of engagements, it is very hard to fathom a span of time as long as this. In order to texture and comprehend comfortably and gratify this difficulty, a pertinent model comes to rescue. Take a human being similar to us standing by our side and place his parent behind him, followed by the parent's parent and so on. We imagine going back in time. *Homo sapiens* are thought to be about 100,000 years old. This will give us an idea about the enormity and true scale of evolution. Our present friend will lead a queue as long as about 5 kilometres and an ancestry of about 5000 generations. Other species of our ancestors such as Homo habilis who are thought to be the first to use the tools are much older—about 2 million years old, they will be about 100 kilometres away. As we get back further and further in time the picture steadily gets hazy. An early outpost was the existence of species which lived about 3 to 4 million years ago, who will be found at the end of the queue stretching about 195 kilometres from our friend leading the queue, standing by our side. This could be Lucy, classified under species Austraopithecus afarensis. *Orrorin* will find the place far behind, last in a queue stretching about 300 kilometres. With this new entry, the process of evolution becomes staggeringly complex. This incredibly long journey makes the puzzle more and more intricate, forcing the inquisitive and analytical faculty of the present human brain to undertake more and more exploration and research activities.

There is an unresolved problem that confronts the scientists in the face, probably one among many. It has been quite intriguing that some of the fossil and other pieces of evidence could not be logically placed in their proper place. It has not been possible to club them with other accepted and established species in the family history of humans either. An account of these cryptic specimens is as follows:

- ✓ *Homo floresiensis,* the Hobbit: In the island of Flores, in 2004 was discovered a fossil, named Homo floresiensis, also Hobbit, as it was about 3.5 ft. in height with an oversized foot, walked upright, used tools and hunted for food. It resembled Australopithecines.

- ✓ *Homo naledi*: Discovered in the year 2013 in a remotely located and inaccessible place in South Africa. It looked like a mixture of Australopithecines and recent, thus regarded to belong to genus Homo.

- ✓ *Red Deer Cave People*: Disappeared around 40,000 years ago, considered the most recent known prehistoric population ever discovered. It was considered separate from *Homo sapiens* and as a separate species by a school of scientists and became extinct without contributing to the gene pool of modern humans.

- ✓ *Penghu Man:* The fossils were dredged from the sea in 2008 in Penghu island west coast, Taiwan. Fossil of archaic human mandible was recovered along with elephant fossils. It was considered to be similar to *Homo erectus* and was named *Homo tsaichangenesis.* The fossil resembled the *Gigantopithecus blacki,* the largest Ape of all time.

- ✓ *Homo georgicus*: It was also named as Dmanisi Man. Some scientists considered it to be placed between *Homo habilis* and *Homo erectus* while others assigned it the status of a separate species. The findings could be taken to suggest that early man did not come out of Africa as seven species but actually a single *Homo erectus* with varying looks.

- ✓ *Denisovans:* It was discovered from Denisova cave in Altai mountains of Siberia and named *Homo denisova.* They are thought to be cousins of *Neanderthals* but their genome studies showed that it contained snippets of DNA that seemed to belong to yet-unknown human species. Interestingly, recent research showed that Tibetans have actually inherited the genes of *Denisovans*.

- ✓ *Boskop Man:* A skull fossil discovered in Boskop, South Africa in 1910, was named *Homo capensis*. It had a small face and larger brain size of 1980 cm$^3$, than 1400 cm$^3$ of ours. After repeated long debates, in 1950 *Boskop Man* was largely settled as a variation *of Homo sapiens*. But the controversy remains alive.

- ✓ *Java man:* It was accepted as Hominid. Dubious gave the hypothesis completely different from the theory accepted and endorsed by almost all the schools of scientists. He propounded that humans were closely related to gibbons and Asia was the cradle of humanity and not Africa. He named it Java Man, after the place of its discovery. Java Man was an upright-walking ape-man and was a link between apes and humans. There have been disputes and debates on *Homo rudolfensis* fossils discovered in Kenya in 1972 and 2012. The other debated issues, continuing unabated even at present, included the direct ancestry of *Neanderthals* assigned to *Homo heidelbergensis* and the practice of *cannibalism* adopted by *Neanderthals*.

The most pertinent of the queries that invoke curiosity while they are still unanswered—who of the lot was our evolutionary ancestor, *Homo heidelbergensis* or some other species whether interbreeding took place between the *Homo sapiens* and *Homo neanderthalensis* and if it did how much; and what is the future of our species so far as evolution of species is concerned. We are genetically different from our *Homo sapiens* ancestors who lived during the period from 10,000 to 20,000 years ago. It is very likely that the rate of evolution for our societies has continuously accelerated since the end of last Ice Age, i.e. about 10,000 years ago. The puzzle as to our real ancestor, identified on the basis of reliable pieces of evidence remains unresolved. The period from about 3 million years ago to 3000 BCE is known as prehistory while the period after that is called history. Based on the advancement of metallurgy and tools, Prehistory is divided into a three-age system. They are Stone Age and Bronze Age followed by Iron Age. The Stone Age is further divided into Palaeolithic Age or Old Stone Age, the Mesolithic Ageron Age or the Middle Stone Age and *Neolithic* Age or New Stone Age. This classification signifies different stages of advancement. The *Neolithic* Age is considered to have begun around 12000 years ago, ending around 3500 BCE when the civilisation started to rise.

The history of modern human evolution may be traced back from the "Early Stone Age", also recorded as prehistory and begins with Palaeolithic Era, marked by primitive technology and culture. This was closely followed by *Neolithic* Age and further by Agricultural Revolution between 8000 and 5000 BCE, in the Cradle of Civilisation, also called the Fertile Crescent and includes the land stretching from River Nile, its delta in Egypt to the Tigris and Euphrates in Iraq. The area encompasses several other countries and their parts such as Egypt, Iraq, Syria, Kuwait, Israel, Palestine, Cyprus, Lebanon, Kuwait, Jordan, the southeastern fringe of Turkey and the western fringe of Iran.

The Old World history comprises "antiquity", the ancient ages up to 475 CE, followed by middle ages, also called the postclassical era from 5$^{th}$ to 15$^{th}$ century and included Islamic golden age from about 750 CE to about 1258 CE, followed by the renaissance that started in Italy and subsequently spread over almost the entire Europe between 14$^{th}$ and 17$^{th}$ century. This period is also referred to as the modern period. The timeline since prehistoric period is graphically exhibited here.

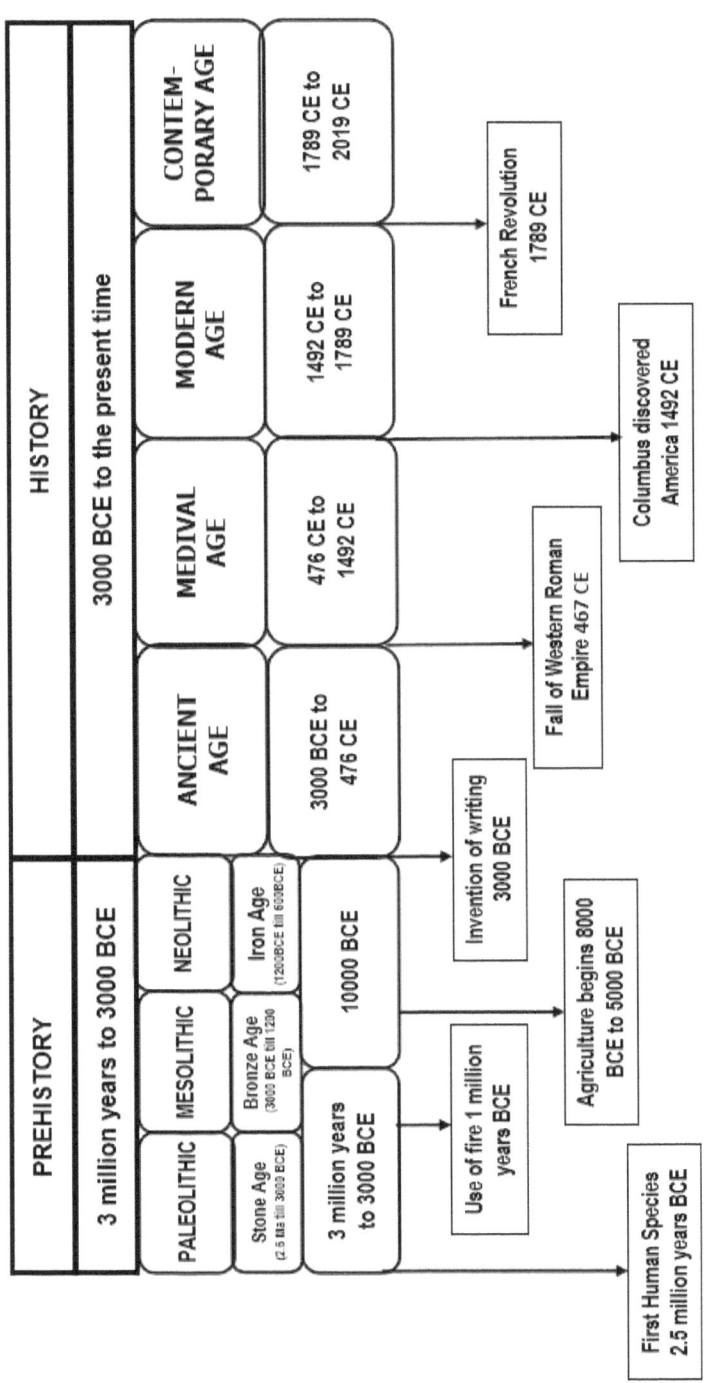

Timeline of Periodic Classification Since Prehistoric Period

*Palaeolithic period* ended with major advancements in the human societies in the form of animal domestication and adoption of agriculture. *Mesolithic period* succeeded the Palaeolithic age. The human societies rapidly got adapted to the global warming to a great extent, which marked the beginning of our present interglacial period about 9600 BCE ago. Nevertheless, their main occupation remained hunting. fishing and gathering until the farmers and breeders took over about 6000 BCE.

The *Neolithic period* was heralded by complete replacement of nomadic life by sedentary one. There was a marked improvement in living standard. Improved stone tools by finer chiselling, use of axes by polishing and making of pottery were some redeeming practices that were adopted by humans of the *Neolithic* period. It was between 6000 BCE and 2200 BCE.

The prehistory is closely followed by the epoch of Protohistory— the metal age. It is characterised by the use of significant technological advances together with the development of metallurgy, mainly composed of copper and tin. The period was from about 3000 BCE to 1200 BCE.

The period from 800 BCE to the end of the first century heralded the making and extensive use of iron in various walks of life. In other words, the period from 1500 BCE to 800 BCE is depicted as the *Iron Age,* followed by the steel age which continued till 1945 while metals and minerals remained in focus.

The period that began in 476 CE and ended around the year 1492 CE is referred to as Middle Age. This period was followed by Early Modern Age starting at the end of 15th and continuing till the beginning of the 17th century.

The period which now followed is known as the Late Modern Period. This period starts from 19th century and continues to the present day. On the contrary, some historians place its beginning in

1789, which also happens to be the end of the Napoleonic period. This age is characterised by the happenings of unprecedented magnitude in the fields of industrialisation, globalisation, political revolution, nationalism, colonisation and wars. It also saw tremendous growth in agriculture, medicine, nationalism and demographic polarisation.

Some great voyagers in the developmental history of humans, such as Marco Polo to present Afghanistan and China, Vasco da Gama to Africa and India, Columbus to the New World added sheen to the planet and helped shaping it the way we see it today. The modern gold rushes leading to the formation of fine modern human settlements like California, Nome in Alaska, South Africa, Australia and Klondike region of the Yukon in north-western Canada—all had minerals, other natural resources and industrial ventures in focus that provided the major incentives for such adventures.

In modern times, the amenities and services invented—and thus available for living a comfortable, contented life—are immense, far more than what the human society had a few decades back, incredibly abundant compared to about a century back; yet, the craving for more lingers unabated. There is a marked reduction in the share of individual beings in the natural material and minerals of many kinds due to various reasons, population increase being one of them. In fact, the production of goods, substantially wasteful and the consumption system adopted by the human society the world over, forces the demands for goods and services to rise immensely. Overexploitation results in the depletion of resources which in turn brings about an increase in the cost of raw materials, energy and minerals and other naturally occurring materials. This generates more waste and causes more pollution. Further in chain is the increase in emission of global greenhouse gases (GHG), resulting from deforestation, degradation of land and irreparable damage to the biodiversity. The development lobby in human intellect vehemently differs from this. Truly it is a vicious circle. The planet warms up from the North Pole to the South Pole. During the $20^{th}$ century average global temperature has increased by about $0.9^0$ C. In order to feel its impact, one need not wait for the far-flung future; it can be felt even now. The phenomena like environmental degradation and global warming bring about climate change. This includes not only rising average temperature, but also causes extreme weather events, shifting wildlife populations and habitats, rise in the sea-level and many other similar disastrous events. By their actions, the human population immensely contributes to their occurrence and increase in their intensity. The climate change plunders the environment.

The population has grown enormously and the environment that prevails is continuously subjecting us to a new natural selection pressure. We are evolving being ruthlessly influenced by natural selection and survival of the fittest. The evolution continues, we sometimes win over the microorganisms and sometimes concede defeat to them resulting into an outbreak of pandemics. Where are we moving to? This is a fascinating query, not only exceedingly baffling but also impossible to answer because of the innumerable unknown factors. We relentlessly continue to evolve until we reach the point of extinction. Why extinction? Just because we tend to distort the precious environment we live in, which selflessly nurses millions and millions of living beings present. In the context ahead we attempt to briefly understand the environment, its vagaries, the *Carrying capacity* of the earth, the commodities that nature supplies free quite liberally, the climate change and similar convolutions. Before delving deeper into the intricacies of such matters, we attempt to understand the fundamental vagaries of nature and other critical factors.

# Chapter 4

# Environment, Its Elementary Concepts and the Veritable Concepts of Free Commodities, *Carrying Capacity* and Sustainable Development

The environment is an intricate phenomenon. It literally means the surroundings; the surrounding objects. The components of the environment broadly include soil, water, air, land, landscape and living creatures. To put it more simply, anything and everything that we experience in nature, the likes of air, water, temperature, moisture, soil, flora and fauna together with microorganisms jointly or singly, their mutual dependence on each other and the non-living matters around form the environment. *The Environment (Protection) Act, 1986* in India, however, defines the environment to include water, air, land and the inter-relationship which exists among and between air, water, land and human beings, other living creatures, plant, microorganism and property. None of them can live singly and in complete isolation. They all exist in closely-knit fabrics and are mutually dependent upon each other. The relationship between all these factors is a copybook example of symbiosis. Environment to more elite shall mean a combination of biotic components of biosphere, hydrosphere, lithosphere and atmosphere in a combined form.

The relationship among these factors when elaborated, forms some important environmental concepts. They are mainly;

i. Living organisms adapt themselves to their environment universally. The past ones did so meticulously and survived only until they were able to do so.

ii. Living organisms are sensitive to the changes in their environment, so were the past ones.

iii. Living organisms possess some inherent characters that impart them sufficient flexibility to withstand the ongoing changes in the environment.

iv. Living organisms exhibit a classic case of symbiotic relationship among themselves and with the contemporary environment. This relationship is extended to conjunctive symbiosis in which evidently, thoroughly dissimilar organisms live in close contact with each other.

The planet earth consists of atmosphere, hydrosphere and lithosphere. The earth, its organisms living as well as non-living and the atmosphere, capable of supporting life, are together known as the biosphere. The major components that form the atmosphere include air, water and land. The environment is the result of a combination of many factors. For the sake of convenience, the factors of the environment may be classified under various categories. The prominent ones are referred to here,

1. Climatic factors are a combination of one or all the factors, viz., rainfall, humidity, light, temperature, wind velocity and atmospheric gases. The climatic factors over a region are generally uniform except for local variations, which are referred to as microclimates and are the results of factors related to topography, etc.

2. Edaphic factors include soil, air and minerals, etc.

3. Biotic factors affecting the environment are the activities of rodents, grazing of animals, presence of plant pathogens and epiphytes. In addition, various types of symbiotic relationship like social and nutritive which are disjunctive and—social and nutritive which are conjunctive. These are also some factors to be reckoned with. The nutritive symbiosis aforesaid means both antagonistic and reciprocal type of relationship; and

4. Topographic and Geomorphological factors include the altitude of the place—the steepness of the slope, the stability of the slope, exposure of the slopes to light, wind and the direction of mountain chains, etc. The topographical and morphological factors are capable of modifying climatic and edaphic factors of a place and affect the flora and fauna indirectly.

The components of the environment include soil, water, air, land, landscape and living creatures. The text ahead describes how the environment is polluted by various sources and the related dynamics. Scientifically, the main components that structure the environment of a planet, continent, place or a domain are elaborated ahead.

The atmosphere of the earth is an intricate mixture of various factors and processes. Major components are the air, the water vapours, the suspended particles of sea salts, silicate—rich dust particles, the organic matter and the smoke. Air is a mechanical mixture of gases, not a chemical compound. Mainly it has four gases—nitrogen, oxygen, Argon and carbon-di-oxide, which are its major constituents. They account for about 95-98% by volume. On the basis of temperature, the atmosphere is conveniently divided into certain well-marked layers. There are three relatively warm layers between 50 to 60 kilometres and above up to about 120 km., viz.,

i. *Troposphere* is the lowest layer, where weather phenomenon and atmospheric turbulence are most conspicuous. It contains 75% of the total molecular or gaseous mass of the atmosphere and virtually all the water vapour and aerosols. It exists nearest to the earth's surface. The thickness of *Troposphere* varies at the poles from about 7 to 9 km. and at equator from 16 to 18 km. This layer also varies in height, is thinner in winter when the air is densest. There is a general decrease in temperature with height.

ii. *Stratosphere* is the second major layer, which extends upwards from the *Troposphere* to about 50 km. Much of the total atmospheric ozone occurs in this layer with peak density at approximately 22 km.

iii. The upper atmosphere comprises several layers. They are:

   1. *Mesosphere* exists above *Stratosphere* in which the average temperature decreases to a minimum of about –90°C around 80 km. This region is generally referred to as the mesosphere. In this region 'noctilucent clouds', the night-shining clouds are observed over high altitude in summers.

   2. Thermosphere, above the *Mesosphere* the portion in which the atmospheric densities are extremely low is known as thermosphere, whose lower portion is composed mainly of nitrogen ($N_2$) and oxygen in molecular form ($O_2$) and atomic (O) forms. At greater heights (above 200 km.), however, atomic oxygen predominates over nitrogen ($N_2$ and N).

   3. *Exosphere* starts from about 500 km. Here the atoms of oxygen, hydrogen and helium form the tenuous atmosphere. The gas laws cease to be valid here. *Magnetosphere* starts after the upper limit of the

*Exosphere*. Apparently, they are formed due to trapping by the earth's magnetic field.

4. *Magnetosphere* is the layer above *Exosphere* beyond 200 km. The ionised particles start increasing in frequency from about 200 km. beyond *Exosphere*, where there are only electrons and protons present. They form two bands at about 3000 and 16000 km. above the earth surface, called 'Van Allen radiation belts'. This layer has an extended tail on the side of the earth away from the sun and is compressed towards the sun by the solar wind.

Satellite data shows that the gases are mixed in a remarkably constant proportion up to about 80 kilometres. The ozone ($O_3$) is concentrated between 15 to 35 kilometres in the stratosphere. There is a constant metamorphosis of oxygen to ozone and from ozone back to oxygen by photochemical processes. As a result, about 40 km. above, an approximate equilibrium is maintained but the mixing ratio is maximum at about 35 kilometres. The maximum density occurs between 20 and 25 kilometres. The formation of the universe is thought to have begun about 13.8 billion years ago, the sun, moon, other planets and earth followed. After about 0.5 to 1 billion years of its formation, the earth started to cool rapidly thanks to the massive collision of a mammoth meteorite. The atmosphere formed. Later on, a few hundred million years ago, the atmosphere acquired its present form and composition when extensive vegetation cover originated on the earth. The atmosphere is vital to terrestrial life, as it provides an indispensable shield against the harmful radiation from the sun and its gaseous contents and supports the plant and animal biosphere. Within the atmosphere, a phenomenon called weather exists and it is neither fixed nor unvarying. The *Stratosphere* is the part of the atmosphere, which is being progressively modified by man's activities. The human activities have a direct influence on this part of the atmosphere. A variety of industrial activities, such as the manufacture of cement, steel, chemicals oil, electrical enrgy from coal-based thermal power plants, textile, some agricultural activities and civil construction are particularly hazardous. The canopy of vegetation cover which when intact functions as a screen, the emanates generated by various sources resulting from human activities, freely move up in the air polluting the atmosphere and disturbing the vital balance of gases. Such man-created pollution, particularly the generation of dust, agricultural practices, forest fires, sea sprays and volcanic activity are all responsible for the entry of aerosol into the atmosphere. The fine particles can remain in the *Stratosphere* above the level of the weather process for one to three years. Similarly, the water vapour content in the atmosphere is also adversely affected.

The indiscriminate human activities often cause undesired damage. Several of the practices adopted by humans bring about complete removal of the top and sub-soil, perhaps the most valuable natural resources, formed as a result of a number of extremely intricate natural processes and agencies over the geological times, none-too-easily to be replaced again. The waste material generated in construction, mining and similar activities block the natural streams when thrown indiscriminately in and around them; such material carried by affluent run-off sometimes reaches the agricultural fields unabated and damages the standing crops, diminishing the possibility of future ones. Perhaps the worst, not visible to the human eye, is the change in natural pH, which adversely affects the regeneration of vegetation.

The natural drainage courses like gullies, streams, etc. become scarce; thus the extra drainage capacity is hampered considerably. The standing pools, therefore, start functioning as recharge grounds and as water barriers to the free flow of effluents into the natural depressions, consequently resulting in the rise in static water levels of shallow groundwater bodies. As a result, the water gets polluted.

Moving downwards below the atmosphere, the other layers are hydrosphere and lithosphere. The hydrosphere is an invaluable component of the planet. The importance and significance of the hydrosphere are so incomprehensible that the life on the planet is possible largely due to its richness. Water is so important that life itself may not be possible without its sustained presence in the living as well as non-living world. The extent is that almost all chemical reactions in process use water either as participant, or medium or catalyst.

Water is present on the earth in many forms and occupies different places; water vapours in the atmosphere, snow above the snow line, affluent seepage on the surface of earth and groundwater. The surface as well as subsurface water accounts for the largest chunk of various forms of water that is used by the living world, to the extent that life literally depends on it. It is a remarkable gift of nature to the living world on the planet earth. As elaborated earlier also, the scientific quest has proved beyond doubt that the life originated in water, it is thriving in water and would vanish once the water vanishes from the earth. There is a complex inter-dependent system involving the atmosphere, the surface and the subsurface water, collectively and popularly called as the hydrologic cycle. According to a conservative estimate, $1.226 \times 10^{18}$ m$^3$ of water is present on the earth, of which approximately 96.5% is encompassed in the oceans, which is not potable for most of the living beings while remaining about 3.5% only is freshwater fit for consumption of most of the living population including the humans. The oceans are the main source of water, with capacity of immense magnitude, from which the entire water originates and to which it all returns.

The plantation that grew in several decades or maybe in centuries is removed to make way for various developmental projects which could only be regrown in a long period and with efforts of much greater magnitude. The groundwater-table gets adversely affected and the process continues through the entire life of developmental projects. Substantial damage is caused to the surface run-off of water in the cluster of activities also, as the stream courses are clogged by heavy silt material generated. The topsoil that forms over centuries through a combination of various agencies working as a closely-knit natural process is lost forever. The soil erosion intensifies as the vegetation is removed. The natural pH of the soil is also disturbed. The exploitative activities invariably cause a dent in some parallel subsidiary event that moves along. Impact of exploitation of natural resources from mother earth is colossal. This arena of commercial activity leaves a trail of destruction behind and requires a distinct analysis. None of the measurement system known to the human brain is proficient to measure it in any unit, probably due to its enormity. We will come across many such instances which were partly or entirely responsible for extermination of some highly-advanced ancient civilisations in our narrative ahead.

The consequences resulting from indiscriminate exploitation of natural resources in terms of what happens to various segments of the environment are summarised below:

i. Deforestation in the areas, i.e., the loss of valuable vegetation cover resulting in the possibility of enhancement of soil erosion, shortage of oxygen availability and increase in temperature;

ii. The loss of top and sub-soil which when removed, adversely affect, the fertility of the land;

iii. Adverse effect on the groundwater-table, the replenishment of aquifers is adversely affected in the terrains devoid of any vegetation, facing intense erosion. Such regions are completely dismembered of aqueducts leading to aquifers. As a result, the effluent discharge of rainwater is increased leaving the water-table partly or very poorly recharged. This also increases the salinity of remaining groundwater;

iv. Due to increased discharge of rainwater passing through the terrains, disturbed by various activities such as mineral mining, the local drainage system is polluted, which on joining the main drainage feature, affects it also;

v. The frequency of landslides increases substantially as a combined result of factors as stated above;

vi. The erosion of soil is enhanced, which hampers the next-generation line of plantations;

vii. The agricultural lands are affected by silt and the fine material released by various activities. It also clogs the surface water channels;

viii. The disturbance caused to the farming lands, adversely affects the well-balanced pH and diminishes the regenerative qualities of soil, etc.;

ix. The disturbance caused to the floral and faunal population makes life more difficult;

x. The heavy earth-moving machines (HEMM) cause problems of noise, vibration and the release of noxious gases in the atmosphere;

xi. The aesthetic damage caused to the landscape reduces its recreational value; &

xii. The disturbance and displacement caused to flora and fauna destabilise the wildlife food balance.

By seeking to understand the impact of human activities such as indiscriminate and hasty exploitation of depleting natural resources, on the atmosphere, its weather and climate, mankind can hope to forecast their vagaries and to a certain extent modify the extreme weather events—gales, tornadoes, hailstorms and floods. There are many factors that cause changes in the atmosphere. These activities bring about fortunes to the progress in human living standards but at the cost of an irreparable loss to the atmosphere, its weather system and the climate.

Conceptually, there are some *"environmental commodities"* which nature has gifted to various living beings on the planet earth. As we have seen in the preceding part, the life did not evolve until the earth was filled with oxygen for breathing. The vegetation on the earth's surface, complements the release of oxygen by absorbing the most powerful greenhouse gas, carbon dioxide, an unwanted gas for humans and other living beings. Do the humans pay even a penny for breathing oxygen? Who creates this oxygen? Similarly, the lifeline on the planet, water is not manufactured in the factories. A natural phenomenon called water cycle fills the rivers and ponds that in turn ensure its supply to the living beings. The population of 7.8 billion or even more across the planet is benefitted without any transaction in any form. The *Stratosphere* enveloping the earth has an ozone ($O_3$) layer that effectively prevents the sun's harmful ultraviolet rays from reaching the earth. An interesting facet will be to calculate the money which such commodities can be exchanged for. Such concepts act as a stark eye-opener for the entire human community the world over. Some highly committed effort with conscious mental makeup will probably create a situation that is needed desperately. Recently, a team of economists and ecologists attempted to estimate the dollar value of 17 worldwide ecosystem services that are not parts of the normally measured money economy—such as water filtration by wetlands, air filtration by plants, nitrogen-fixing by soil microbes, and the like. Their estimate is that these 17 services that we take for granted and get for free without any transaction in terms of money, are worth something like twice the entire gross global product, and they believe that this estimate is probably far too low. Such alarming interpretations come from a brilliant piece of work by Robert Costanza et al. (1997), an American/Australian ecological economist and Professor of Public Policy at *Crawford School of Public Policy at the Australian National University*. The terminology may be incomprehensive, the numbers fuzzy, and non-materialistic, but they are powerful indicators of the magnitude of our dependence on nature.

Yet another concept that invokes our thought process, probably confronted in the past also by exceedingly developed human societies and civilisations— *"concept of carrying capacity"*. On investigating the matter like environment, one has to take the entire globe as one unit, with no geographical boundaries. It is simply humans, cattles, land, oceans, rivers, air, water, flora and fauna together with surroundings. This brings in the bothersome riddle related to the limits of resources and the capacity of the planet to absorb the users. The most baffling question haunting the environmentalists, equally the novices, is the maximum number of humans and animals which the planet earth can support. This is called the 'carrying capacity'. Broadly *Carrying capacity* may be explained as allowing damage to itself. This underlines the importance of land together with other resources. However, population numbers alone do not inform us of the total burden we place on our environment. Resource use and waste output per person vary immensely from country to country and from society to society, and will probably keep on changing in the future too. For each necessary resource, it is important to determine the burden we are now putting on, as well as the maximum sustainable burden beyond which that resource will degrade.

It has been pointed out on several platforms across the world that local activities are the source for many problems. The local bodies construct, create, operate and many times maintain also the economic, social and environmental infrastructure. The planning process, implementation of local environmental policies and regulatory provisions often receive poor attention thus initiating a chain of problems. The understanding of *Carrying capacity* provides an operational framework for planners for sustainable development. The *Carrying capacity* is determined by the single vital resource in its least availability. The modern human society, however, depends on many ecological and economic resources for its survival. It may be concluded reasonably that the ultimate aim of probing the specifics of *Carrying capacity* is to produce desired goods and services from limited resource base without damaging one or many environmental segments beyond a certain specific level.

The *Carrying capacity* expresses varied meanings for different aspects. *Carrying capacity* is a term used by ecologists to describe the maximum number of animals of a given species that a habitat can support indefinitely, without permanently degrading the environment. According to Odum 1997, the *Carrying capacity* for human society can be defined as the maximum rate of resource consumption and waste discharges that can be sustained indefinitely in a defined planning region without progressively impairing bio-productivity and ecological integrity.

*Carrying capacity* is an ecological concept that expresses the relationship between a population and the natural environment on which it depends for ongoing sustenance. The concept assumes limits on the number of individuals that can be supported at a given level of consumption without degrading the environment and, therefore, reducing future carrying capacity. That is, *Carrying capacity* addresses long-term sustainability. The concept of *Carrying capacity* in the modern context refers to the number of humans who can be supported without degrading the natural, cultural and social environment. Exceeding the human *Carrying capacity* implies impairing the environment's ability to sustain the desired quality of life over a long term. The concept of *Carrying capacity* is displayed ahead diagrammatically:

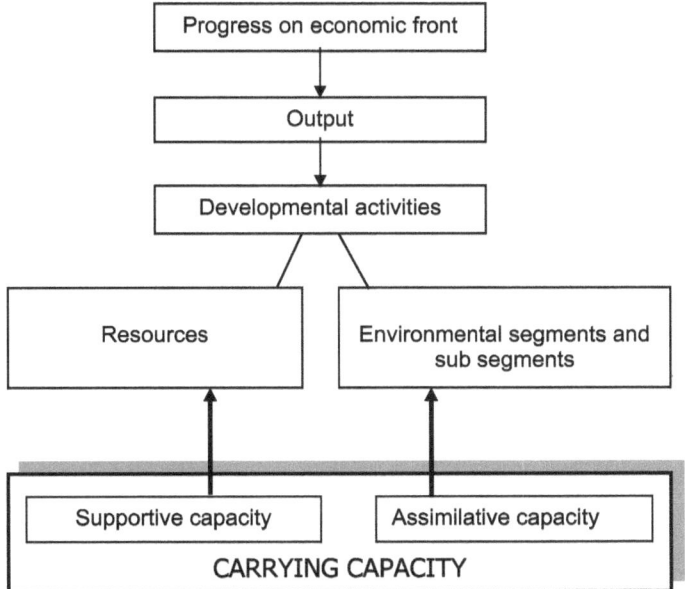

The environmental *Carrying capacity* is a venerable, even though hypothetical concept. The concept of *Carrying capacity* is widely discounted, in part because it is fluid and virtually unquantifiable. One group of scientists advocates for technology sternly, saying that technology's potential is unlimited. Technological optimists typically reject scientific warnings that no substitutes exist for topsoil, fresh water, clean air, and the "free services" of many species. The standard answer to relevant pieces of evidence that non-renewable resources are being depleted, or a renewable one degraded, is that, if a resource becomes "scarce" or pollution too detrimental and environment too degraded, prices will rise sufficiently to call forth either substitutes or innovative technology that overcomes the problem. The *Carrying capacity* of a region or area can be measured. Some methods frequently used for measuring the *Carrying capacity* are,

i. Resource-production relations: It may be defined as the capacity of available resources to sustain rates of resources used in production.

ii. Resource-residuals relations: It is the capacity of the environment media to assimilate wastes and residuals from production and consumption at acceptable quality levels.

iii. Infrastructure-congestion relations: It is the capacity of infrastructure resources (distribution and delivery systems) to handle the flow of goods and services—resources used in production.

iv. Production-social relations: It is the capacity of both resources and production outputs to provide an acceptable quality of life.

Lately, the human society has become development-oriented. Invariably, the development is oriented to more and more comfortable lifestyle of human beings. This obliterated the method of governance also in certain societies and largely the countries. Some of the most powerful economies completely favour the capitalistic social set-up leading to the creation of wider gaps between the affluent and downtrodden strata of their societies. It is a situation which inflicts a serious blow to the measures needed for safeguarding the environment on the planet. This led to the development of the highly acclaimed concept of 'sustainable development'. It is the development permitted to the extent 'just necessary' with no harsh and detrimental effect on the surroundings whatsoever.

In 1987 the Brutland Commission coined the term 'sustainable development', a new paradigm that is defined as, a development that meets the needs of the present without compromising the ability of future generations to meet their needs (NR Can, 1997:4). Critical assumptions in support of this prototype are the role of technological advances and the high level of substitution between various forms of capital. The capital here may include financial, social, human, manufactured, intellectual, and natural capital. It also assumes that the total stock of assets remains intact and that they are completely interchangeable. This assumption was adopted and verified by

Robert Solow, the 1987 Nobel Prize winner for economics, whose work in 1974 was among the earliest research to assess the impact of depletion on intergenerational equity. He demonstrated that the maximum-minimum criterion was a reasonable criterion for intertemporal planning decisions. By introducing exhaustible resources into his analysis he also found that the elasticity of substitution between natural resources and labour-and-capital goods was less than unity. He concluded that earlier generations are entitled to draw down the pool of natural resources provided they add to the stock of reproducible capital. Investigations into intergenerational equity undertaken by Mikesell (1989) also suggest that mineral resources should be regarded as a form of a social capital asset, the value of which should be preserved for future generations even though the minerals may be extracted and consumed by the present generation. This applies to all the non-renewable natural resources. In line with the arguments of both Solow and Mikesell, Blignaut and Hassan (2001) have stated that it is necessary to compensate future generations with fixed stocks as a result of consumption of the natural assets by reinvesting part of their value (resource rents and any form of revenue generated) in other forms of capital assets that can provide the same stream of benefits in the future. This does not mean that the asset base of natural resources is not reduced, but the ability to generate a stream of income into the future remains intact. Mikesell is of the opinion that annually saving and reinvesting an amount equivalent to the net revenue from the sale of natural products can achieve sustainability. This again assumes the substitutability of reproducible capital for wasting assets as they form an intricate commodity.

Risk evaluation, reduction, and management are major elements of sustainable development. Evaluation and management of risk can be achieved by developing or reinforcing the human, scientific, institutional, legal and information capacity of a country. Critical developmental factors are exceedingly significant in many ways. The factors of development, which are most pertinently responsible for the criticalities, also have an impact on the environment. Likewise, various sub-segments of the environment affect these developmental factors. These factors may include land use, water use, water quality, air quality, socioeconomics of the area, public health, ecology and resource position of various natural resources including the minerals or rocks being taken up for exploitation. The most responsible one should be identified. In fact, critical developmental factors, as the term implies are often necessary no matter what trail of damage is left behind. It is, therefore, obvious that they have a greater impact on several sub-segments of the environment. These factors need meticulous identification followed by elaborate topic-by-topic documentation. A set of mitigating measures specifically applicable to the respective development factor is worked out, taking local conditions into the consideration. Based on the set of mitigating measures, all put together, the strategy is to be designed, workable under the existing set-up.

# Chapter 5

# Empathy of the Climate Change, Global Warming and Environmental Damage

"Climate change" and "global warming", though not interchangeable literally, are often used in lieu of each other as they virtually carry the same insinuation, yet have distinct meanings, just like the terms "weather" and "climate" are also sometimes confused, though these events mean two different events dissimilar in spatial and timescale implications. Global warming and climate change are not synonymous at all. The atmospheric conditions occurring in a small area over a short period denote the weather in that area. The events like rain, snow, clouds, winds or thunderstorms are certain examples. Climate, on the contrary, refers to the long-term regional or even a global average of temperature, humidity and rainfall patterns over seasons, years or decades or even longer. In plainspoken words, global warming is just one symptom of the much larger problem of climate change. The variations in the mean state and other characteristics of climate "on all spatial and temporal scales beyond individual weather events" is called the climate variability. In other words, climate variability is a combination of various factors, the ones not caused systematically but at random times. Two international fora, namely the *Intergovernmental Panel on Climate Change (IPCC)* which involves several countries and the *United Nations Framework Convention on Climate Change (UNFCCC)*, address the issues related to climate change.

Let's get into the basics of these phenomena and the responsible causes one after the other. The term *climatic change* was proposed by the World Meteorological Organisation (WMO) in 1966, to encompass all forms of climatic variability on time-scales longer than a decade, but regardless of causes. Climate change is a long-term change in the average weather patterns which defines earth's local, regional and global climates. These ranges of weather patterns signify a certain set of effects. Climate change refers to the variations that persist for a longer period of time, typically decades or more. In order to focus on anthropogenic causes, the term initially coined *"climatic change"* was replaced by *"climate change"* in the 1970s. It had the same meaning and implication but with a broader arena of causative sources. Lately the term 'climate change' is used both for a technical description of the process as well as a noun to describe the problem. The climate change encompasses global warming but refers to the broader range of changes taking place on the planet. These include rise in sea levels, shrinkage of glaciers and ice sheets, acceleration of ice melting in Greenland, Antarctica and the Arctic and shift in flowers/plants blooming times. Change in earth's climate is observed since the early 20$^{th}$ century. This is primarily attributed to human activities, particularly the running of automobiles by fossil fuel burning, which increases heat-trapping greenhouse gas levels in Earth's atmosphere, raising earth's average surface temperature.

Some factors called *climate forcing* or *"forcing mechanism"* are the factors that play a significant role in shaping the climate. These factors include atmosphere, oceans, variation in earth's orbit, variation in solar radiation, variation in reflectivity of the continents, also called albedo, mountain building, continental drift and most crucial of them all, the changes in the concentration of greenhouse gases. In addition, the rate at which the sun releases energy and the rate at which it is lost to space determine the equilibrium temperature and climate of the earth. There are certain agencies, which carry the energy released by the sun around the planet to affect the climates of different regions. They are ocean currents, winds, etc. Climate change points out the changes caused by human activities as well as by natural causes. Natural processes can also contribute to climate change, including internal variability that includes cyclical ocean patterns like *El Nino*, *La Nina*, the Pacific Decadal oscillation and external forcings such as volcanic eruption, changes in the sun's energy output and variation in Earth's orbit.

The temperature produced due to activities of the human population is commonly referred to as global warming. Global warming may be understood as long-term heating of Earth's climate system observed during a period of 50 years from 1850 to 1900, the pre-industrial revolution period. Global warming is thought to have been created by human activities, primarily the burning of fossil fuel. The period before the onset of the industrial period also witnessed an increase in the temperature of the earth's surface globally brought by human activities. Based on the observation currently recorded, Berkeley Earth, a non-profit research outfit, based in Berkeley, California, made a shocking revelation. The global temperature will increase by about $1.5^0C$ or $2.7^0 F$ above the average recorded in five decades from years 1850 to 1900. This increase is thought to occur by the year 2035. At the same rate, the global temperature of the earth's surface will experience an increase of $2^0C$ or $3.6^0 F$ by the year 2060. This is a frightening situation. The most potent factor responsible for such precarious trend will probably result from human activities. Along with currently increasing average global temperature, some parts of earth may actually get colder while others may get warmer. It is for this reason; the temperature of the earth is talked about as average global temperature. When the heating and agitation of the atmosphere is caused by the greenhouse gas effect, it is complemented by an increase in unpredictability of weather and climate. It also brings about a dramatic increase in the severity, scale and frequency of storms, droughts, wildfires, and extreme temperatures. Global warming can be irreversible and the warming can reach the point when all forms of life will face extinction including humanity. According to one estimate, the increase in average temperature in the range of $2.2^0C$ to $4^0C$ above the average temperature of the pre-industrial period is considered as associated with triggering and marking the beginning of irreversible global warming process. An increase in average global temperature levels by $5^0C$ to $6^0C$ leading to the extinction of all the life on the planet and eventual loss of our atmosphere are contemplated by the environmental scientists.

Over the last century, the average temperature on the planet has increased by about $0.6^0C$. A record of temperature during 19th and 20th century shows that the warming trend of about $0.5^0C$ began in the early 19th century, peaked in the decade of forties and a decrease is observed in the next 30 years, up to the decade of seventies. An increase is observed from 1976 onwards. In the mid-20th century, the increase observed is primarily due to an increase in greenhouse gas emission resulting from human activities.

Long-term and periodic climate change is a phenomenon that needs a mention and is dependent on the emissions of heat from the sun. Sun emits a lot of heat, a billionth of which reaches the top of the atmosphere, called insolation and keeps the earth sufficiently warm. This solar radiation increases when the earth moves close to the sun and decreases otherwise. The types of change in the earth's orbit cause some prominent variations which include the shift in Earth's orbit from near-elliptical to near-circular; the change in the tilt of the earth's axis of rotation; and the wobbling of the Earth's axis of rotation. They bring about the changes in the climate.

The medieval warm period from 1200 BCE to 1000 BCE, the historic time and short Ice Age from 1800 BCE to 1550 BCE were some prominent periods which can be referred to as short-term climate changes. Some factors aided the occurrence of climate change though they were not adept at doing it on their own. These factors mainly included volcanic eruptions, sunspot variations, changes in greenhouse concentration in the atmosphere, changes in large scale ocean current conveyor belt, etc. The intensity of solar radiations is an important factor, other than the Earth's orbital parameters—both of which are considered to be responsible for short-term climate change for about the last 10,000 years. This results from the development or decay of sunspots, produced by sun flaring—the magnetic activity inside the sun, which takes place from time to time. The period of *"solar maximum"* is followed by *"solar minimum"*. As specified earlier, they denote the sun's intense and lean or inactive phases.

The study of Paleoclimate is equally significant and is always a great help in deciphering what is likely to happen in future on this front. Briefly, this is done by four types of data, namely, historical data; instrumentally recorded data; satellite data; and proxy records of climatic parameters, which include tree ring records and coral growth patterns, etc. In some of the categories of data type, the data was obtained only in the period when there were no thermometers.

The human and cattle populations contribute immensely to the pollution on the earth. Their activities are the major contributor. Some of the major sources of emission of greenhouse gases are,

- ✓ Industrial activities: Cement manufacture—Emission of $CO_2$; Fertiliser manufacture—Emission of $N_2O$; Nitric Acid manufacture—$N_2O$; Refrigeration—CFCs; Aerosol spray propellant and cleaning solutions—CFCs;
- ✓ Transport and Biofuel burning: $N_2O$, $CO_2$, Ground-level ozone and other greenhouse gases (GHG);
- ✓ Energy: Coal and oil-fired generators and thermal power plants—Emission of $CO_2$;
- ✓ Mineral exploitation & deployment of HEMM: Coal—Emission of $CH_4$ and other noxious gas;

- ✓ Agriculture activities such as rice cultivation—Emission of $CH_4$ & application of chemical fertilisers;

- ✓ Waste management: Anaerobic decomposition of manure from the waste of domestic animals, like cows, etc. Industrial and municipal wastewater in sewages— Emission of $CH_4$ and other hazardous gases;

- ✓ Land use Change & Forestry: Wood logging from forests, conversion of grasslands into farming lands—Emission of $CO_2$. The USA, Australia, Europe and some Asian countries are the highest contributor of $CO_2$ and other noxious gases by fossil fuel burning and industrial practices.

The adverse consequences of global warming, undoubtedly, are immense to the extent that they might leave the planet completely without any ambience whatsoever. Scary anticipations of global warming might leave a human brain completely dumbfounded. Some of the major consequences of global warming are:

- ✓ Melting of glaciers and ice sheets;
- ✓ Rise in sea-levels resulting in inundation of coastal areas;
- ✓ Increase in storms;
- ✓ Increase in southwest monsoon incidence;
- ✓ Adverse impact on water resources and unbearable decline in water availability;
- ✓ Fluctuation in crop yields and changes in irrigation needs;
- ✓ Changes in forest composition, geographic range and productivity of land;
- ✓ Increase in levels of weather-related mortality, infectious diseases and respiratory ailments;
- ✓ Conspicuous changes in floral and faunal population.

It is now resolved that with phenomena like climate change and global warming, the human world will probably be completely changed upside down. The question if the life will still exist on the planet having been adversely affected by the aftermath of these phenomena, after about a century from now is rather tough to answer. The economics, health, cultural, social and ethical fall out apart, demographic commotion is likely to be encountered. On the contrary, the advent of colder climate in decades and centuries ahead will throw the entire planet, its enormous diversity and vibrant life completely out of gear. The human activities are probably hastening the arrival of doom. Shockingly, we will see ahead that some of the ancient and highly developed civilisations were completely devastated and were eradicated due to these or similar factors.

During the fascinating stride through the evolution of humans after acquiring the anatomy and the DNAs of *Homo sapiens*, living through a legacy of about 100,000 years and 5000 odd generations they also obtained manifold jump in the arena of culture, governance and overall living standards. Several highly developed and advanced civilisations prospered in various places on the planet. These civilisations evolved, adopted a steady growth pattern advancing gradually and at one point in time disappeared unceremoniously. A look into the history of last about 5000 years and beyond, the life of *Homo sapiens* unveiled a very fascinating account. After the times of Indus-Saraswati Civilisation or maybe the Sumer Civilisation, the record is almost authentic and is relied upon. Some highly-revered Hindu scriptures are considered to be about 5000 or more years old. It is this period that will be oriented to see the growth and downfall of various civilisations which were instrumental in shaping the present form of the human world together with the possible causes for their extermination—whether environmental, something other than that or both.

The human evolution, as complex as it is, becomes intricate when an attempt to link the older times with the present is made.

Civilisation and culture of modern times owe a lot in terms of their roots to earlier civilisations who in turn had emerged after millions of years of progress made by humans on the evolutionary path. Gradually the practice of living in groups with mutual cooperation, understanding and dependence among constituent members of the assemblage, was adopted. Slowly the groups became larger and developed into societies who further in time advanced into civilisations. Such transformations, quite a few in count, resulting from specific human mentality and psychology, have probably been acquired from the closest cousins, the members of genus Pan—chimpanzees and bonobos. It forms an interesting subject matter of research for historians and anthropologists. It began, progressed and has come a long way from an age which had no defined means of communication, and scavenging and hunting-gathering were the primary sources of food and life. With time, agriculture took over from foraging and hunting, the animals were domesticated and societies matured which eventually led to what we are today.

Some prominent civilisations formed, progressed, flourished and got eradicated after maturing and developing enormously. The question arises as to how and why such bright and progressive people living in harmony could not survive the odds that might have struck. Probably it was some natural catastrophe of such colossal dimensions which proved beyond human endurance. Such problems do occur even today and are often irreparable and beyond the capability of present humans, in spite of being loaded with the power of machine and technology. Where does the environmental degradation and collapse play the critical role? Environmental degradation leading to climate change, cumulatively leading to collapse occurs when societies overshoot the *Carrying capacity* of their environment. This ecological collapse theory, which has been the cause of great concern, the subject matter of meritorious study, points an accusing finger to matters like excessive deforestation, water pollution, soil degradation and loss of biodiversity sometimes leading to the extinction of species. Was the environmental degradation, climate change and the resulting adversities responsible for the fall of such exceptionally-developed civilisations? Was the fine balance of sustainability disturbed? We will make an effort to answer these haunting questions in our postulate ahead.

With the purpose to briefly revisit many stopovers during the arduous but fascinating journey you have cherished till now, the portrayal is updated, before the narrative negotiates a major turn. After the appearance of universe, earth and life on earth, the narrative elaborately comprehended that the period witnessed a major event in the evolution of *primates* in addition to numerous others, followed by their bifurcation into *Prosimians* and Anthropoids which took place about 50 million years ago. The evolution of Anthropoids further progressed and about 8 million years ago, another event of divergences began to occur—genus Pan to which chimpanzees and bonobos belong, took one branch, while the *Ardipithecus* group, the occupant of bottom-most position in human family tree broke off from the common ancestors and started to tread the other branch. After gradually evolving into groups such as *Australopithecus* and *Paranthropus*, the group Homo evolved. In due course of time; the world we live in transformed from hunter-gatherers to more organised social hierarchies and institutional governments. The inventions by the early hunter-gatherer groups led to the formation of more organised trade, agriculture and the division of labour started to flourish. This happened when *Homo sapiens* with the largest cranial capacity started to learn its use. It took nearly a million years for evolution from *Neanderthals* to *Homo sapiens* who gradually evolved into the modern human over a period. The process did not stop there. Modern human, the species *Homo sapiens*, progressed to develop a culture of good, hygienic living, better dwelling practices, systematic clothing, more productive agricultural practices, animal domestication, advanced health care systems, poultry by farming and nursing large social and demographic groups. With incredible and sustained advances in science and technology, the human had become equally destructive. Probably with a rather natural course of demographic and social polarisation, various civilisations with specific cultural style paved the way for life and made enormous progress. Control of fire also was a proverbial landmark. Despite the advancements, they were required to move from one place to another looking for new and richer avenues of food. The discovery and adoption of agriculture are, without doubt, the biggest achievements in the history of mankind. They could now stay in one place for a longer time. The dependence on hunting had drastically reduced or had got eliminated from the life of certain groups. All were invariably positive to one thing, common in their agenda —the goal of a better and more sophisticated pattern of life must be achieved. With agricultural practice thus a settled life, it was now possible to feed more people. With more food, the population grew dramatically and so did the scientific research the world over. With the development of settlements, the conflicts between the groups emerged and people felt the necessity of a government to administer the law, protect them from bad and evil people and take the decision for the welfare of societies or the group. This gave birth to the concept of power. With trade starting to flourish, the division of labour became inevitable. The taxation system soon got introduced and flourished. This helped the groups or societies to run the affairs more prosperously. A new era had begun —the human civilisation.

About 5000 or more years ago, people started living on the banks of the river. The abundance of water in the adjoining rivers made agriculture easier. They started to domesticate the animals for food and as marketable commodity. In due course of time, man had started to realise that the source of food and the access to commodities like water were limited. Various groups started to compete with each other, sometimes in fine fettle but often violently. The like-minded groups thus developed into civilisations of specific nature, some of whom grew into mighty power centres in times that followed. The capacity of the resource was restricted but the takers were growing, the calamities of nature were sometimes too harsh and the animal instinct of protecting and capturing those resources was growing continuously. In the late 19th century, the modern thinkers were quick to realise the value of vistas like environment, ecology on the planet, the optimal use, besides the conservation of resources, sustainability, global warming and broadly the climate change. The worry on this account escalated more and more with the passage of time in the 21st century.

This has taken geological epochs, eras, periods and times immemorial to reach this monumental juncture in modern times from antiquity. The apprehensions are different now. It is the economy, welfare, infrastructure in various human life segments, nature, environment and climate change. The study of these abstract and lasting but true events of paramount value occupies the prime position for the living world. As we see, these scientific disciplines seem to have their roots firmly set in the deterministic traditions. The collapse is not a parable now. The mankind had agonised before also, to the extent that centuries-old well-established civilisations were doomed.

The prominent civilisations and societies which formed, progressed and collapsed include, namely, Indus Valley Civilisation, Sumerian Civilisation, the Maya Civilisation and the Rapa Nui society. Some of the other prominent civilisations include ancient Egyptian Civilisation, the Chinese Civilisation, the Ancient Greek Civilisation, the Persia Civilisation, the Roman Civilisation, the *Aztec* Civilisation and the *Inca Civilisation*. Four of these otherwise prominent civilisations, their beginning, progress and collapse happened rather mysteriously and herein we attempt to attribute the possible reasons for their debacle.

# Chapter 6

# Dawn of Civilisation: Indus Valley Civilisation, the Master Traders in Erstwhile Indian Sub-Continent

The inception of Indus Valley Civilisation, also known as Indus-Saraswati Civilisation and Sindhu-Saraswati Civilisation or the River Valley Civilisation is considered as a major event in human history. It progressed in the alluvial planes of River Indus, River Saraswati and other places, proceeded to become one of the biggest establishments in large areas, among all civilisations which we know of in the time span of disciplined human societies. The fertile flood-plains of River Indus in Northwest India and present Pakistan presented an ideal venue for agriculture and dwelling. The flood-plains and alluvium of Ghaggar-Hakra river, which drained into revered and mythical Saraswati, were also inhabited by this civilisation. Three mighty rivers Sutlej, the *Shatadru* in Indian scripture originating from Mount Kailash, *Drishadvati* from Siwalik Hill ranges and the ancient Yamuna from the Himalayas, flowed together along the channel occupied by the River Ghaggar-Hakra. Ghaggar flows mainly in the Indian states of Punjab and Haryana, called Hakra in the Indian state of Rajasthan and Nara in Sindh. Based on the support of satellite imagery studies, it is more or less accepted now that *Shatadru* and *Drishadvati* were the tributaries of Saraswati. The ancient Yamuna originated from Yamunotri from frozen Champasar lake/glacier located on the Kalind Mountain at a height of 4421 meters above the sea-level. After a short stride, River Yamuna is joined at Hanuman Chatti by River Hanuman Ganga which in turn receives its feed from the glacier located on the south side of Bandarpoonch mountain massif.

American writer, vedic researcher and explorer David Frawley, taking the cue from the oldest sacred Hindu scripture the Rig Veda which refers to River Saraswati several times, thinking Saraswati River flowed along Ghagghar-Hakra, named it as "Saraswati Civilisation" or "Indus-Saraswati Civilisation". Geophysical investigation has clarified that Ghagghar-Hakra river system was a monsoon-fed river which later on became perennial about 4000 years ago, when the civilisation also shrank. Ultimately it came to be famously known as Indus Valley Civilisation. It was further opined that the Indus Valley Civilisation essentially developed to become an urban civilisation where the main occupation of the people was trade and commerce. On the contrary, the Vedic Civilisation was rural with people's main occupation being agriculture. They were not the same. Till some time back, many thought that beginning of the civilisation came with Rig Veda but it is now firmly concluded that Indus Valley Civilisation was not Vedic Civilisation and came before Rig Veda.

The place of its discovery, the type site is Harappa and thus it is also named Harappan Civilisation. A more appropriate name, however, would be Sindhu-Saraswati Civilisation or Indus Valley Civilisation. The area inhabited by the populace of this civilisation covered a large span of land, almost triangular in shape, spanning over about 1,299,650 square kilometres. It stood as the largest civilisation, both by the size of the area inhabited as well as the population among all the ancient civilisations. Broadly, the area extended in the north up to Shortgai in present-day Afghanistan; up to Rakhigarhi in the present Indian state of Haryana; up to Somnath in the south-east to the south-west in the present Indian state of Gujarat; and up to Suktagen Dor in present-day Baluchistan province of Pakistan. A population of about 5 million strong, roughly 10% of the world's population at that time nursed the advancement and progress of

Indus Valley Civilisation at its peak time. It was first identified in 1920 in Harappa, located in west Punjab by an Archaeologist Rai Bahadur Daya Ram Sahni, CIE, a protégé of John Marshall, the then Director-General of Archaeological Survey of India in British-ruled country. Later on, Archaeologist Sahni took over as the first Director-General of Archaeological Survey of India. In 1922 Mohenjo-Daro was discovered by R.D Banerji, an officer of the Archaeological Survey of India near Indus River in Sindh region. They are both in present-day Pakistan. The name Mohenjo-Daro invokes the sense of "the mound of the dead". The civilisation spread far and wide from modern-day Afghanistan to Pakistan and from north-west to the middle of India and far beyond. The civilisation was initially dated about 3300 BCE, while some others place it much earlier and support their findings by interpretation of the results obtained from intensive research. This civilisation flourished till about 1300 BCE. The major places where the urban establishments were developed with their present names are, Alamgirpur, in Meerut district, Bargaon, Hulas in Saharanpur, Sanauli, Sothi in Baghpat district, Mandi in Muzaffarnagar district of Uttar Pradesh; Babar Kot in Saurashtra district, Bet Dwarka in Devbhoomi Dwarka district, Bhagatrav in Baruch district, Dholavira, Desalpur, Juni Kuran, Khirasara and Gola Dhoro in Kutch district, Kaj, Kanjetar, in Gir Somnath district, Kuntasi, Rajodi in Rajkot district, Loteshwar in Patan district, Lothal, Rangpur in Ahmedabad district, Malwan, Surat district, Nageshwar, Pabumath, Shikarpur, Surkotada in Kutch district and Vejalka in Botad district of Gujarat state, Balu in Kaithal district, Banawali, Bhirrana in Fatehabad district, Rakhigarhi, Hisar, Siswal in Hisar district, Jognakhera in Kurukshetra district, Mitathal in Bhiwani district and Farman in Rohtak district of Haryana state; Daimabad in Ahmadnagar district of Maharashtra state; Ropar in Rupnagar district of Punjab state; Karanpura, Kalibangan of Hanumangarh district, Baror in Sri Ganganagar district of Rajasthan state in India. The places of Shortugai of Takhar province, Mundigak in Kandhar province of Afghanistan have also been the venue of cities belonging to Indus Valley Civilisation. In what is the present-day Pakistan Sutkagan Dor in Makrana district, Pirak of Sibi district, Pathani Sheri Khan Tarakai in Bannu district, Rehman Dehri in Dera Ismail Khan district of Khyber-Pakhtunkhwa, Pir Shah Jurio, Allahdino in Karachi district, Larkana, Mohenjo-Daro in Larkana district, Kot Diji in Khairpur district, Lakhueen-jo-Daro in Sukkur district, Chanhudaro in Nawabshah district of Sindh province, Pathani Damb, Nausharo, Mehrgarh of Kacchi district, Pathani Damb in Makran, Balakot in Lasbela district of Balochistan province; and Harappa in Sahiwal district are some prominent places which had well-developed cities and urban establishments of Indus Valley Civilisation.

The well-known sites and the largest cities built by such a large and widespread civilisation are Mohenjo-Daro, Harappa, Mehrgarh now in Pakistan and Rakhigarhi, Lothal, Dholavira and Kalibangan in India. These are the places where pieces of evidence are present in all their grandiose, exhibiting the commendable progress touching the pinnacle of social, commercial, agricultural and religious advancement, the civilisation had made before its downfall started, probably gradually. The beginning of Indus Valley Civilisation could also be taken as 7000 BCE to 6000 BCE propounded by some workers on the basis of the site in Mehrgarh in present Pakistan. Additionally, pieces of evidence related to religious activities date back to about 5500 BCE. This site is considered to be an early part of *Neolithic* Age from about 5000 BCE to 3500 BCE. It is thought to have reached the zenith during the Bronze Age, about 3000 BCE to 1200 BCE which is generally deliberated to mark its demise also. The decline started around 1800 BCE. The mature period of the growth of this civilisation was from 2600 BCE to 1900 BCE, also called as Early and Late Harappan period. Remarkably, this period also witnessed the introduction and acceptance of a culture in which the ranks of predecessor and successor were established. It is thought to have lasted for about 2000 years from 3300 BCE to 1200 BCE and would have nursed 35 to 40 generations. Extensive research work, undertaken by the *Indian Institute of Technology, Kharagpur* yielded astonishing results. The researchers used a technique called 'optically stimulated luminescence' used to date the shards of pottery from early mature Harappan time assigning it the age to about 6000 years ago and the cultural levels of pre-Harappan Hakra phase to as far back as 8000 years from now. More scientists joined in, started working with *Physical Research Laboratory, Ahmedabad*, made use of C14 analysis for analysing the remains of domesticated animals to decipher the antiquity and the climatic conditions which existed during the times, the civilisation flourished.

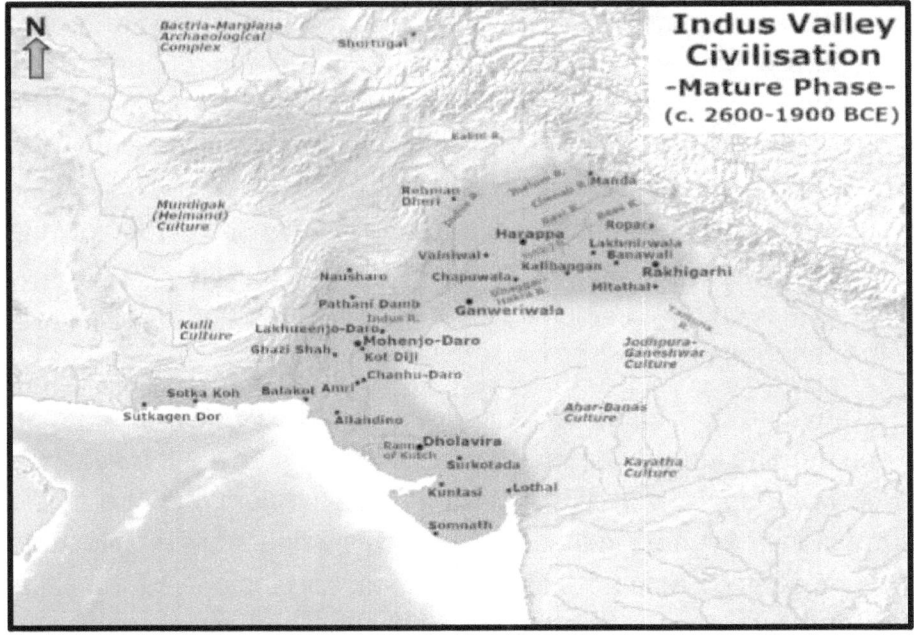

Tentative Location of Inhabitations Developed by Indus Valley Civilisation

The Indus people developed into a highly enterprising society. They practised agriculture as their main occupation. Initially, the farmers established their dwelling units on the river bank, as they relied on river water for household chores, drinking and farming.

There must have been enough rains. The rivers had an enormous impact on their life. To Indus Civilisation people, rivers were the '*The King River*'. The flooding and backwaters of the river brought huge quantities of fertile soil and minerals which ensured good agricultural yields, subsequently enabling the progression of societies and wealth in general. The crops they cultivated included wheat, barley, peas, lentils, other pulses, millets, linseed, mustard, etc. However, no evidence of rice-eating is traced. Did they know the emission of noxious $CH_4$ gas from rice farming? It can only be guessed. The Indus people domesticated animals like cows, goats, sheep, dogs, cats, humped and short-horn cattle, horses, donkeys and probably pigs also. Elephants were probably used for their ivory. Horses and donkeys were used for transport. They ate a variety of food and were not entirely vegetarians. They also ate meat, beef, bird meat, fishes and shellfishes depending upon their abundance in the vicinity. The Indus people were enterprising traders also and relied on it to a great extent. They developed trading links with different civilisations such as Persia, Mesopotamia and China. They were also known to trade in the Arabian Gulf region, central parts of Asia and parts of what is presently Afghanistan. Interestingly, they had trading links with Sumerians in the then Mesopotamia.

They had developed a fine sense of art, architecture and urban planning. The large and well-planned cities with sanitation facilities, large water pools, provisions for community podium, etc. are ample proof of their acumen in these fields. It was a fine art of architectural excellence and demographic governance. The majority of the cities they built were small. As many as 1025 cities and settlements were built, while seven of them were truly large. The Indus Valley people used metallurgy. The realistic metal statuettes are strong pieces of evidence of the expertise possessed by the Indus people. The most famous is the lost-wax casting bronze statue of a slender-limbed dancing girl, adorned with bangles recovered from Mohenjo-Daro. A near classical treatment of the human shape discovered from

Harappa as seen in the statue of a dancer apparently a male and a red Jasper male torso are some more examples of the craftsmanship of people in Indus Civilisation.

Impression of Kalibangan cylinder seal *K-65;* Indus Valley Civilisation
*Two warriors spearing each other, and holding the hand of a lady, while a woman dressed as a tiger stands nearby*

The Indus people built houses using baked bricks of the uniform ratio of 1:2:4, made up of mud and wood. Mohenjo-Daro and Harappa were the prominent and well-developed cities. The cities built in Dholavira and Lothal were planning marvels. The Indus Valley Civilisation people's religion was Polytheistic-Hinduism, Buddhism and Jainism. They had a complete understanding of the environment. The Pipal tree, *ficus religiosa* was sacred to Indus Valley people and they worshipped it as done by the Hindus even on this day. Similarly, they had compassion for animals too and they worshipped animals like bull and tiger, etc. The Indus Valley people also worshipped the sun, fire and water. Probably, they were aware of their value to the very life on earth. Proof gathered from seals discovered from the related archaeological sites have figures inscribed of them which closely resemble the Gods Shiv and Rudra. Lord Shiva as Pashupati Nath was a popularly worshipped deity.

The religious cult of the Indus people is supported by many reliable pieces of evidence. The Indus people had immense reverence for Mother Goddess who symbolised trees, plants and nature. The Mother Goddess was thought to be a symbol of fertility also. They extensively used Hindu auspicious symbol of "Swastik", which they probably designed originally. Magic, charms and sacrifices formed a part of the belief of Indus Valley people. They showed full respect for their dead. They were either buried or cremated. The burial was done with several household items and ornaments, etc. On choosing to cremate, the ashes were collected in earthen urns probably for disposal after some ceremonies. Their belief in life after death was strong and is demonstrated by these rituals. These practices are akin to the system adopted by the Hindu religion, which continues with the modern-day Hindus also. The Indus Valley Civilisation people were rather liberal in the social system with no religious dogmas and in this, they were opposite to other contemporary civilisations which probably lived in other parts of the globe. We do not exactly know anything about their method of governance, yet they probably had no system of kings, chieftains or any such authority in their system. The civilisation was split into several domains each governed by some major cities. Further, it was made into villages, cities and regional councils, all governed by the supreme Harappan Council. Therefore, it could have been a kind of federation instead of Indus kingdom.

The Indus Valley people gave some pioneering innovations to mankind which shows their outstanding mental capabilities. They developed and were probably the first to use precise measurement techniques. They were pioneers in building artificial dockyard. They pioneered to make buttons and developed great craftsmanship skills. The Indus Valley people also pioneered the construction of step-wells and dug-wells and were the first to use them extensively. They started the use of seals for identification of various status holders. They built large public bathing premises. They are said to govern their families on the matriarchal form. Their script was intricate, comprising symbols to depict various objects. This script is quite a large corpus of symbols and is written from right to left. In spite of continuous endeavours, the scholars have not been able to decipher the script, probably due to the absence of any bilingual script. Truly the Indus Valley Civilisation was way ahead of their times. In addition to sanitation, they had developed the concept of household toilets also. Interestingly, they monopolised the mining of some gemstones like *Lapis Lazuli*. The trading colony of Indus Valley Civilisation in *Shortgai,* Afghanistan, located in the northernmost end of Indus Valley Civilisation regime was established around 2000 BCE on the banks of *Amu Dariya,* the *Oxus River* near the *Lapis Lazuli* mines located in the *Sar-I Sang* in *Shortugai* and *Badakhshan* province in northeast Afghanistan. These mines are dated about 6000 years old. This semi-precious gemstone is prized since the ancient times for its intense blue colour *(Firozi)* and brought in incredibly large fortunes to Indus Valley people through its export. They obtained other semi-precious stones from Bhagatrav also, bordering the south-east edge of the area covered by civilisation, located in the present Gujarat state of India. Bhagatrav also made a significant contribution to the revenue which probably shaped the base for the creation of Dholavira like marvels of engineering and urban planning. However, little could be read out of their script till now resulting in complete inability to understand the institutions and system adopted for governance of their societies.

In spite of extensive research conducted for a long time to ascertain the period when Indus Valley Civilisation started taking shape and began its journey, reached the stage of maturity and culminated to disappear completely or ended by getting assimilated into some other contemporary civilisation, success seems far away. We do not know exactly what had actually happened except that a well-groomed, economically sustained and stable civilisation went into oblivion. When did this actually happen and what might have caused this? The scientists are trying to solve such intricate riddles about one of the oldest civilisations on the planet. By working out the causes for the plight of Indus people, we might be able to adopt precautionary measures for the safety and prosperity of our future generations. However, the Indus Valley Civilisation flourished on the earth in the range of time span from 5500 BCE to about 1200 BCE, which is supported by pieces of evidence, thus is largely agreed upon by various schools of scientists. However, the opinions vary significantly on other aspects. This civilisation reached maturity and evolved into a fully grown human society between 3300 BCE to about 1300 BCE.

The decline of Indus Valley Civilisation, the largest and most developed of all the Bronze Age civilisations has been an unresolved riddle, vaguely answered even on this day. Truly, this situation has come up mainly due to fact that authentic records of Indus-Saraswati Civilisation were systematically destroyed and burnt by the regime of later rulers. Whatever little was saved, posed another problem—the language which accounts as yet another reason for ambiguity. Another possible reason is probably the limited studies of certain traits of civilisation, the script they used in particular, which still remains indecipherable. Initially, it was thought to be sudden and total degeneration. Later on, resulting from more studies, it was concluded that the decline was not dramatic and sudden. One school of scientists even argue in favour of continuity and survival of the civilisation in certain other areas. In fact, some settlements did deteriorate profusely, particularly in areas like Sindh, whereas there has been a remarkable growth in Ganges-Yamuna region in the west and in present Gujarat in the southwest. The transitional phase is sometimes marked by a shift from urban settlements to villages. However, the pioneering distinctive features of Indus people did not remain in place, probably disappearing forever. This interpretation is based on its reference in Mesopotamian literature in which there is a reference of *"Meluha".* This is how the Sumer Civilisation called the Indus Valley people As already stated elsewhere, the Indus inhabitants were trade counterparts of the people of Sumer Civilisation. The reference of Meluha did not appear after 1900 BCE in Sumer records.

In order to explain the causes behind the decline of Indus Civilisation, a rational view is necessary. It was popularly propounded by a British Archaeologist Mortimer Wheeler that an ethnolinguistic group comprising Indo-European people, who were skilled fighters also, migrated to the Indian sub-continent, fought with Indus Valley people and drove them out which ultimately led to the complete and sudden decline of the Civilisation. This theory also received the colour of Dravidians, projected as the original inhabitants and Aryans, the invaders without assigning an element of time and sequence of the events. On the contrary, Indus Valley inhabitants had evolved into a social network of harmonious and conflict-less population. The Aryan invasion theory became popular probably because of the very limited research undertaken until then. Some of the skeletons, particularly the one from Rakhigarhi did not show any of Aryan gene—the R1a1, which implies that people from steppes of central Asia migrated to Indian sub-continent after the decline of Indus Valley Civilisation. This theory was mainly publicised when the British ruled over what was then the undivided India, quite in the original shape. This theory, therefore, stands rubbished by subsequent workers and is debunked. Another possible cause ascribed to the civilisation's decline and subsequently complete obliteration is a natural catastrophe on a large scale that eradicated the civilisation. As is most conspicuous, the Indus cities and establishments were located hundreds of kilometres apart. For instance, Shortugai, now in Afghanistan was about 1000 km. away from Mohenjo-Daro. There is hardly any piece of evidence to indicate the occurrence of a natural catastrophe in the region. Assigning one single reason for the Indus-like diverse civilisation, inhabiting such a large area does not appear a plausible notion and is thus not accepted. With this, yet another cause or a combination of causes has to be found out to solve the mystery.

The foremost fact that needs to be clarified is whether the eradication of civilisation was gradual. Although one region might have collapsed due to some cause, the other kept on with life for a noticeable period even afterwards. Indus Valley Civilisation like other contemporary civilisations developing in the distant corners of the world was riverine in nature. The River Valley Civilisation or river culture is the one that dwells in the areas of a river basin and draws sustenance from the river. They are based on the resources provided by these rivers, which include irrigation in agriculture, drinking, household chores, fishing, transport and to cater to the needs of domesticated cattle. Indus Valley Civilisation was a classic example of river valley civilisations. Larger and more advanced cities were located either on the banks and basin region of rivers Indus or Ghaggar-Hakra river system and their tributaries. They include *Mohenjo-Daro, Harappa, Ganreiwala, Kalibangan, Rakhigarhi, Ropar, Dholavira* and *Lothal*. The sense of adoration and the importance of water in their life were immense. This can be seen in the methods adopted in the structure of their houses in Mohenjo-Daro, Harappa and design for water management system together with the drainage system in the city of Dholavira. The city of Dholavira was positioned to develop between the two rivers, bordered on the north by River Mansar and River Manhar bordered the city on the south. These rivers were joined by tanks setting up a model of probably the oldest application of the concept of river linking. Dholavira in present Khadirbet Island in the Kutch Desert Wildlife Sanctuary of Kutch district in Gujarat was probably a busy port in the Arabian Sea during the period of Indus Valley Civilisation. Spread over an area of 47 hectares or 120 acres, built in a quadrangular shape, the city of *Dholavira* had roads at right angles. The city was planned with due consideration of mathematical accuracy and 'Vaastu'. The exact ratio of 5:4 among the walls was adhered to in the entire city.

Lothal, another Indus Valley Civilisation city was located between the rivers Sabarmati and its tributary Bhogavo in Saurashtra region. Though the sea was about 20 km. away at that time, probably the boats from the Gulf of Cambay could have sailed right up to the spot close to the habitation. In the present Indian states of Rajasthan and Haryana, the Indus Valley cities developed along the course of River Ghaggar-Hakra, which included Kalibangan, Rakhigarhi, Banawali, Birrhana, etc. This cluster of locations was probably their bread-basket. This leads us to orient our thought to the calamity, if any related to water, and more specifically to rivers which could be assigned as the cause for the annihilation of this great civilisation. The streets covered with a thick bed of silt transported sediments and tumbled buildings and houses point to inundation in Mohenjo-Daro, Dholavira and Kalibangan cities. Mohenjo-Daro experienced anomalous flooding at least three times and the beds of transported sediments identified are the proofs. Such repeated catastrophic floods hampered the normal irrigation practices leading in turn to the economic decline in some of the Indus Valley's flourishing cities. Some scientists, however, disagree with this idea particularly due to the presence of massive silt deposition that according to them cannot result from the normal behaviour of a river. River Indus flows in an active seismic zone. Some kind of uplifting of the plain, blocking the passage of Indus water channel might have led to the inundation of the natural banks in the adjacent area followed by submergence of cities like Mohenjo-Daro. This too was contradicted on the ground that huge discharge of Indus water could have easily breached such raised levels of earth. Later the silt deposit in Mohenjo-Daro was identified as a combination of *eolian sand*, *silt* and disintegrated mud, unlikely a result of flooding. Urban Plan of Dholavira city also disregards this idea The tanks built by the planners connecting the two rivers bordering the city on two sides and both finally emptying in the Arabian Sea, could have easily maintained the discharge so as to minimise the possibility of any flooding even in case of disproportionate rains. The tanks were built by maintaining the gradient meticulously thus ensuring proper discharge of water. The city also had a very effective drainage system built using the natural gradient. The analysis of skeletal remains discovered from Mohenjo-Daro does not support any massacre, as no signs of trauma are visible. There are phenotypic differences from other Harappan population as shown by the skeletons, supporting the presence of a heterogeneous population. Phenotypic patterns shown by the skeletons are found to be unchanged before or after the decline of Indus people. This clearly indicates that no new groups migrated into the region.

The ethnicity of the population of Indus Valley Civilisation has remained a subject matter of continuous debate since their existence came to the knowledge of modern human society in the year 1920. The most widely spread hypothesis stated that the original inhabitants were people with dark skin, named Dravidians. Invaders, migrating from various places as far as Europe, named Aryans drove off the Dravidians and inhabited the green pastures along the banks of these rivers. This popularly came to be known as the Aryan invasion theory. In order to substantiate this opinion, a finger is pointed to the potentially divisive demographic set-up in present India. In plainer words the Dravidians were driven off to the far end of the country, to occupy the southern part while the invaders occupied the northern part of what India was then, something that still testifies, a thought born out of demographic set-up, the contrast in people's skin colour and linguistics. We attempt to delve into the roots of this pretended and unsubstantiated theory. The support for this theory was heralded by British rule over India who used the theory of Max Mueller, among others, whose interpretations of ancient holy writings of Hindus, including Rigveda, hypothesised that the chariot warlords were the invaders from the north. This was unscrupulous and British amplified this theory and used their colonial influence to spread it across the world. Their narrative was that the Indian sub-continent was always inhabited by a less civilised lot and later on came to be dominated by more decent and civilised ruling class from the north, just like the Britishers themselves. This was a classic piece of wrong and as to the science, undeniably a piece of pseudo-science. As if it was not enough, controversial but popular then, a racist occultist, philosopher and author, Ukrainian Helena Petrovna Blavatsky, also known as Madame Blavatsky added more toxicity and propounded that the Germans are also descendants of these so-called Aryans. Interestingly, some German occultists, including Hitler, happily agreed to this. It looked as though in ancient times the Indian land was romped about by a *Nordic-esque race*, something that is nothing more than unsubstantiated nonsense. Alternatively, it was also thought that some migration of the groups of people took place into the areas inhabited by Indus Valley Civilisation. This migration might have taken place in phases and the migrants were apparently absorbed in the original river valley societies. There is no piece of evidence to support the idea of any type of large scale encounters. An almost complete absence of any weaponry in the Indus Valley establishments refutes the possibility of any war or war-like situation in any of the excavated sites. It is evident from the preceding part of the narrative that they were master architectural planners, built well-planned cities, efficient drainage system, all based on sophisticated engineering concepts, managed the water resources and were very proficient traders having trade links with other contemporary civilisations located far away. It thus becomes hard to concede to the idea that the influx of migrants happened and drove off the people inhabiting the land. Instead, they might have happily got absorbed with the population already there. It is pertinent to bring about the excerpts of what is opined by the exceedingly respected Dr. David Frawley, also known as Pandit Vamdev Shastri, a Padma Bhushan awardee by the Govt. of India, otherwise a Catholic and a fellow of *American Institute of Vedic Studies*. He is of the opinion that the Saraswati was a great river more than 5000 years ago and dried up around 4000 years ago. It finds reference in ancient literature like the Rigved, in the later Vedic literature also and even in Mahabharat, the popular Indian multi-faceted-epic which incorporates the God's own statement, Shri *Bhagwad Gita*, in which Lord Krishna is said to have participated as a normal human, the charioteer. This clearly means that these people knew about the revered Saraswati through its different stages. With this very well in place, the hypothesis that Aryan invasion took place around 1500 BCE can be contentedly thrown in the waste paper basket. That these people flourished on the land of Indian sub-continent 5000 years ago, is established beyond doubt by their knowledge of the continent's river system, even of the one that ceased to be seen on the surface, the revered Saraswati. It can be conclusively taken as proven, according to Dr. Frawley, alias Pandit Vamdev that the hypothesis of Aryan invasion of the Indian sub-continent who destabilised the Indus Valley inhabitants is unfounded and wrong. For this, David Frawley's monumental work *Gods, Sages and Kings; Vedic Secrets of Ancient Civilisation* has provided exceedingly coherent sustenance. Frawley is candid and indisputably authoritative when he says, that India was an "advanced, spiritual and poetic culture, not a group of primitive nomads."

The economy of Indus Valley Civilisation was agrarian. The farming during the Indus Valley period was entirely dependent on the hydrographic networks and ultimately on rains. The civilisation prospered in the nuclei around the course of Indus and Ghaggar-Hakra river systems, together with *Manhar, Mansar,* Sabarmati and *Bhogavo* Rivers. The river dynamics has best been studied by application of satellite imagery and study of aerial photographs. The theory, that the course of Indus River might have shifted to cause a situation that led to the collapse of civilisation, was later disagreed upon assigning in support of this conclusion the enterprising nature of the inhabitants. The paleochannels clearly show the pattern adopted in the past. Drying up of the rivers appears to be more convincing. The upper reaches of Indus, the main inhabiting venue of civilisation has been studied extensively, including the courses of rivers like Ravi and Beas. Near Harappa, the river Beas is represented by a dry bed with river placers in the channel. Many relict channels have also been identified. Several Indus settlements developed around the course of Beas which is ample proof that Beas was an active river during the times.

Based on the study of paleochannels and the drainage pattern analysis in the region, it is concluded that Sutlej changed its course several times towards the north during time range from 5000 to 3000 BCE. General gradient and river's own environmental setting played a critical role in this. River Saraswati dried up, virtually disappearing completely, and this had a major adverse impact on Indus Valley settlements. This led to a general collapse of civilisation due to some kind of 'domino effect' across the establishments. Migration and relocation undoubtedly took place on a large scale, the collapse of Indus Valley Civilisation cannot be accredited to any single factor. At least one fact appears to have been agreed upon by most of the schools of scientists that it was a gradual process involving many factors in spite of some advocating in favour of sudden demise of the civilisation. This gave future explorers an opportunity to identify factors related to climate change and global warming, broadly environmental disaster. A catastrophe like an earthquake can bring and hasten the collapse but leaves a trail of solid pieces of evidence behind, all of which are conspicuous by their complete absence in establishments of Indus Valley Civilisation.

An interesting method of working out the possible causes of decline and abandonment of the civilisation has been suggested by an American archaeologist Walter Fairservis. He advocated for anthropogenic causes responsible for dismantling the ecological balance, leading to decline. The population around Mohenjo-Daro, the largest city of Indus Civilisation of humans and cattle was estimated. This Indus city inhabited a human population of 41250 persons, cattle population of 8700 and land around 22715 acres was under cultivation. The population increased and so did the demands in due course of time and kept on increasing. The growing demands of the population led to the depletion of forest, food and sources of energy. Later on, some scientists raised questions on the number of cattle which did not include the numbers deployed in lift irrigation and thus the number was underestimated. It also required a network of villages and the agriculture labour force from such villages. These numbers must not have been constant. Certain aspects, as pointed out by the scientists, reasonably stand against this and refuse to accept the validity of the argument. Such schools of workers argue that the computation of the population number is based on a very dicey piece of evidence and therefore in serious doubt. Truly such calculations and subsequently the interpretations necessitate the understanding of the relationship with nearby village gentry for which no plausible calculation is done. However, principally this method is applicable, as it very strongly points out to the sustainable balance among various commodity providers, the nature in this case and the user inhabitants. Trading network and links with potential traders and contemporary civilisations also need to be understood. Yet another argument on the same line is pushed forward which states that the Indus flood-plains have been fertile and have catered to the needs of the teeming population over the last two millennia. This inflicts a doubt on the soil exhaustion hypothesis, even though over-cultivation and over-grazing might have destabilised the symbiotic relationship and the ecological balance among various factors in the region. The floods resulting from overexploitation of resources and extent of alterations of the landscape caused by human activities might have triggered the inundation of riverbanks leading to devastating floods. This ultimately induced the salinity in the soil demurring the future crops in the region, which reduced the yield year by year. Subsequently, the deterioration in town planning and living standards is reflected ending up in total decline. This was possible if the climate remained much the same. According to Shereen F. Ratnagar, an Indian archaeologist, environmental degradation appears a more plausible cause than a natural catastrophe. Apparently, anything whose effect that lasts longer has to have a more devastating effect on a civilisation so large and versatile. She also points out that the consumption of wood charcoal affecting the forest in the region as the fuel source for copper metallurgy could have been a possible cause. She explains that as per the estimate of Horne in order to process 5 kg. of copper, about 100 kg. of wood charcoal is needed to be burnt requiring 700 kg. of fresh wood or the product of felling of 25 to 30 trees. This was an additional factor causing stress on the resources ultimately by their reduction.

The drinking water mostly obtainable from as many as 700 dug-wells in Mohenjo-Daro was the only source to cater to the inhabitants. The salinity and contamination of dug-wells cannot be ruled out in the groundwater-table through local replenishment and the floods. Some towns and cities of Indus Valley Civilisation might have encountered the problems, aggravating beyond forbearance, caused by floods which might have been the cause for hastening the demise. Trade, its growth and expansion in Ganges plain and present-day Gujarat led to the over expansion. The political and economic system had probably become too big. Textural reference to *'Meluha'*, as the Indus region was named by Mesopotamians had started to decrease substantially. The Indus materials like gem Carnelian and varieties of Quartz Chalcedony, *Lapis Lazuli*, gold and ivory items had probably exhausted which were traded to Mesopotamians and coastal areas of Arabian Peninsula only up to 1900 BCE. The trading focus had actually shifted from Sindh to coastal settlements in *Omana Ras-al-jinz*, *Bahrain* and *Gujarat* indicating the continuity of trade. The interaction with *Bactria Margiana Archaeological Complex (BMAC)* had started increasing. The establishments in Sotkakoh and other places were closed. Even the *Lapis Lazuli* trade centre in Shortugai, Afghanistan was seeing much less Indus Civilisation material, replaced by BMAC material. These pieces of evidence do point out that trade was vital to the civilisation and trade's potential decrease must have contributed to its demise.

The ambiguity in working out the details of various walks of life of Indus Civilisation people is mainly due to non-availability of any written records. Their script is intricate to the extent that scientists have not been able to decipher it as yet. The records must have been prepared by such an intelligent, elite and versatile civilisation. Such records prepared directly by them or got prepared by a team must have existed. These records might have been stored safely, most probably in some institution. Mughals systematically destroyed the universities and together with it the valuable and authenticated literature also. The biggest and most coveted university, the Nalanda University was burnt by Mamluk invader and Turkish leader Bakhtiyar Khilji. It was under active fire for 30 days. Later on, the ones that escaped Muslim rulers, fell in the hands of Britishers, who distorted them by redrafting the historical themes and disseminated their version.

According to Ratnagar, "Collapse means the end of an integrated and complex social, economic and political system, carrying with it a decline stratification, erosion of economic specialisation, the eclipse of regulatory institutions and the flow of information, the city life that embodies the sophistication of the civilisation and ultimately the monuments and art production." Although some traits and elements related to rural technologies and peasant knowledge that was passed down to the generations did survive, Indus Valley Civilisation in the form of "a political economy with its institutions of dominance, its economic networks, its interrelationships, dependencies and intellectual norms, in other words, a state system and the cultural circumstances in which it had flowered" did come to an end, as stated by Ratnagar. It was probably an environmental disaster stretched to cover its various factors, to the extent of imbalance in the sustainable system that was the cause for its demise which led the people to either shift to greener pastures on Ganges basin and maybe further to the present-day Gujarat, discarding the identity. Different sites faced different problems, the ones close to River Saraswati dissipated when the river dried, Shortugai on the northernmost end of Indus Valley Civilisation area declined due to the fall in the trade of *Lapis Lazuli* and Dholavira due to climate change resulting in drying up of Manhar and Mansar Rivers bordering it which was the root cause. When the city-states collapsed, obviously the trading establishments followed or vice versa. The Indus Valley inhabitants taught the modern humanity several lessons that were not known before and gave irrevocable traits. Even after such a magnitude of advancement on social, scientific and technological fronts the civilisation perished carrying some unresolved mysteries with it.

The fact is that the rivers whose banks provided the dwelling venues and where the Indus Valley people thrived for thousands of years have changed their courses as it happens even today with various rivers. The erosion and breaching of river banks is a prevalent phenomenon which causes havoc to a large number of dwellings and population. The problem is both ways—either when rivers inundate the dwelling places on the banks or when they move away from establishments. The complete drying up of River Saraswati brought off a major setback which consequently hampered its tributaries also. A study conducted by the geologists and geophysicists in *Indian Institute of Technology-Kharagpur*, West Bengal state, India yielded astonishing results and validated the idea that the Indus Valley Civilisation was confronted by a long drought in different places probably at different times. The sediment samples drawn from Tso Moriri lake located in the Ladakh region of Jammu and Kashmir revealed that around 4350 years ago, the summer rains started to weaken. In fact, the rains played truant in the entire north-west Himalayas for a long period of about 900 years. This part of the Himalayan region fed most of the rivers in the northern part of the peninsula—directly or through the supporting surface drainage system. The surface effluent seepage would have become non-existent, adversely affecting the replenishment of groundwater-table. Indus Valley inhabitants, the inventors of dug well, must have been hard-pressed even for drinking water. Consequently, the rivers which acted as the source of water for irrigating their crops would have dried up. Such shortage as appeared during the period of 900 long years eventually drove them off, the otherwise enterprising and industrious people towards east and south where the rains were better, regions water-rich and the pastures greener. They might have mingled with other contemporary, small and inconspicuous groups and societies. In a broader sense, the drought over a long period is certainly indicative of climate change. This is what might have happened to a part of the sub-continent and brought off the extermination of a highly developed, intelligent and ingenious people, the creators of one of the oldest civilisations on the planet. The theories and analyses presented above are generally based on certain assumptions supported by a near-to-exact scintilla of logic, historical support and rarely on hearsay, particularly when the subject matter is as old as 3300 BCE, quite likely far beyond it.

## Chapter 7

# Dawn of Civilisation: Sumerian in the 'Cradle of Human Civilisation', Did Not Disappear Suddenly

The name Sumer owes its origin to the language *Akkadian* of the north of Mesopotamia, which literally means "land of the civilised kings". The southernmost region of modern-day Iraq and Kuwait was the ancient Mesopotamia, also named Sumer. This area is generally considered the *'cradle of civilisation'* and is called the 'Fertile Crescent' also. Almost contemporary to Egyptians in the African continent and Indus-Saraswati Civilisation in Indian subcontinent, another civilisation arose in the southern Mesopotamia. This was the enterprising Sumerian Civilisation. Geographically, Mesopotamia lay between Tigris and the Euphrates rivers, in the area that later became Babylonia and is now southern Iraq from around Baghdad to the Persian Gulf. The Sumerians called themselves "the black-headed people" and the land they occupied was just "the land" or "the land of black-headed people". Sumer in the Biblical Book of Genesis is known as Shinar. The city of Uruk is considered the oldest fully developed city in the world. They themselves believed that it was Eridu, in modern Arabic Tell Abu Shaharain, where the order was first established and civilisation began.

The region of Sumer did not receive the Sumerians as their first settlers. People of unknown origin whom the archaeologists have termed the Ubaid people, taking the cue from the excavated mound of al-Ubaid where the recovered artefacts endorsed their existence, were probably the first to inhabit the region of the Euphrates River. These people—whoever they were, had evolved from hunter-gatherer society to agrarian society about 5000 BCE. The articles recovered in the process of excavation included painted pottery, clay artefacts, figurines, stone tools like knives, hoe and sickles, etc. They were actually the first to propagate civilisation in the region. When did the Sumerians enter, consequently replacing the Ubaid, is not exactly known at all. Probably for a certain period both were present in their respective places.

The Sumerians contributed enormously to the evolution of disciplined societies and fought for resources, a prominent animal instinct in humans that continues even today. The kings of Ur, Ur-Nammu from 2047 BCE to 2030 BCE and Shulgi from 2029 BCE to 1982 BCE more or less pursued the line of today's welfare state. The Sumer Civilisation made cultural advancement and overall progress as the goal of their administration and attempted to maintain harmony which ensured the required platform of progress on the fronts of art and preliminary technology. During and before the period of Ur dynasty, the inventions made included tools, some technological headway and application. Certain path-breaking concepts consolidated the Sumerian's place in history as the creators of civilisation known to us and adopted as constructive even today. Samuel Noah Kramer, a world-renowned *assyrilogist* listed 39 'firsts' from the region in his book *'History Begins at Sumer'*. The Sumerians are credited for their initiation for starting the schools, the proverbs, the sayings, the messiahs, the first love songs, the Noah and flood stories, the funeral chants and the tales of a dying and resurrected God. Sumerians were also the first to promulgate the moral phrases across the globe. The Sumerians are considered the pioneers in many such things that were essential and are equally relevant today.

Sometimes around 5400 BCE, the city of Eridu was founded by people named Ubaid. The trading outpost, thought to be those of the Ubaid people, Godin Tepe in the valley of Kangavar in Kermanshah province is a settled Sumer region, inhabited by Ubaid people around 5000 BCE. The Sumer Civilisation is known to have adopted the practice of burying their dead with respect. Ubaid people flourished in the region from 5000 to about 4100 BCE, whereas the period from about 5000 BCE to about 1750 BCE was the booming period of Sumer Civilisation. The Sumerians had deep-rooted belief in God and built their first temple in 4500 BCE and simultaneously founded the city of Uruk. The religion trusted by Sumerians was anthropomorphic polytheism. Each city had its own precise god in human form. The God *An* to Sumerians represented heaven, their God *Enki* was the healer and friend to humans, God *Enlil* controlled the spirits, *Inanna* was the god of love and war, sun god to them was *Utu* and moon god was named *Sin*. The first written evidence of religion in Sumerian cuneiform came into reality in 3500 BCE. The Sumerian population prospered energetically and they developed writing in 3600 BCE at Uruk. The first instance of written language in Sumerians came up in 3200 BCE. The first empire in Mesopotamia was established under the first King, named Eannutum, who laid the foundation of the first dynasty in 2500 BCE. The literature in Sumerian began almost during the same period. The first code of laws by Urukagina, the king of Lagash was introduced in 2350 BCE. During the period from 2150 BCE to1400 BCE, the Sumerian epic *Gilgamesh* was written on clay tablets. The list of Sumerian kings was created in 2100 BCE in the reign of Utu-Hegal at Uruk. Later the king prevailed to rule the *Akkadian* cities also during about 2055 BCE to 2047 BCE. Ur-Nammu dynasty came to reign over Sumer from 2047 BCE to 2030 BCE. The last period of Sumerian history during the period of king Ur III from 2047 BCE to 1750 BCE, is also known as Sumerian Renaissance. This period is marked with the construction of the great wall of Uruk, still standing partly intact. This was a remarkable period of exceptional advances in culture, covering virtually every aspect of civilised human life. Later another king Shulgi of Ur built his great wall during 2038 BCE. Yet another prominent landmark in Sumerian history was the creation of the code of Hammurabi considered as one of the earliest codes of law in probably the entire civilised world. During 1750 BCE Elamite invasion and Amorite migration marked the end of Sumerian Civilisation. Later on, in about 1100 BCE the details of creations of Sumerians, called 'Enuma Elish' are made from the much older text referred to frequently now. The Sumerian culture comprised a group of city-states. At its peak around 2800 BCE, the very first true city, Uruk was inhabited by about 78000 to 85000 people and was enclosed within about 10-kilometre-long wall serving as a defence structure. In all, there were about 30 cities of different sizes housing varied numbers of people. The other cities which had flourished include Eridu, Ur, Nippur, Lagash, Kish, Uruk, Bad-Tibira, Larsa, Katallu, Nina, Urukag, Girsu, Umma, Adab, Kissura, Isin, Pazuris-Darjan, Marad, Dilbat, Borsippa, Babylon, Kazallu, Kid-nun, Kutha, Akshak, Agade, Sippar, Neribtum, Tutub, Eshnunna Der, Borahshi, Bit Bunakki, Mari and Rapiqurn. These were the cities of Sumerian Civilisation built and developed in the area between the rivers Tigris and Euphrates. The cities of Ur and Eridu occupied the southeastern end of the region while Mari bordered the north-western end.

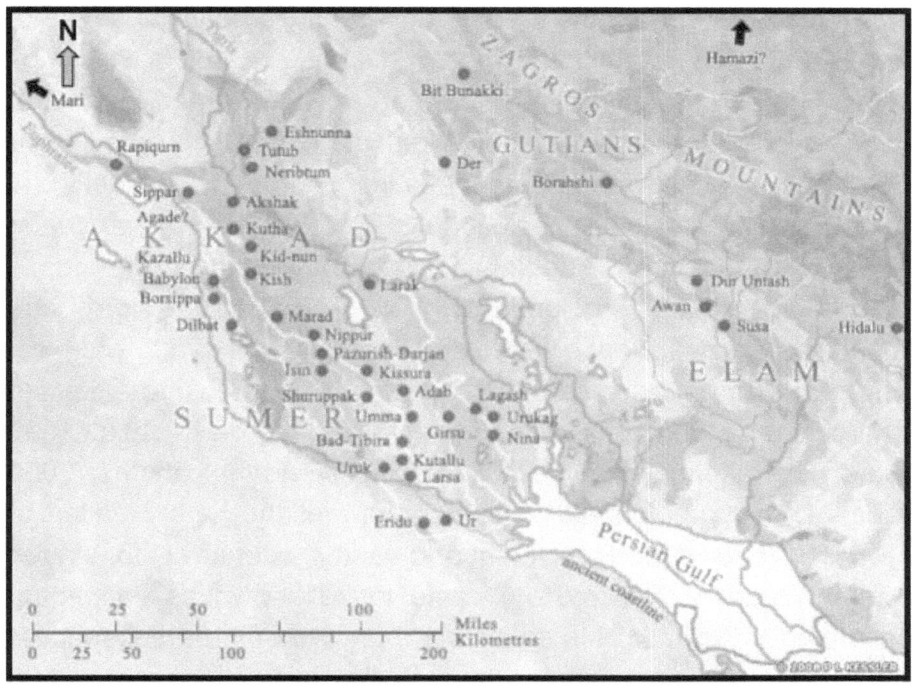

Tentative Location of Inhabitations Developed by Ancient Sumerian

One of the most prominent feathers in the crown of the Sumerians was structuring and writing of the first legal code which became the precedent in future and is vastly known as *Hammurabi of Babylon*. The third dynasty of Ur king was recognised as the unified force among the Sumerians and became a *Patrimonial state*. As a result, the monarch was given the status of a highly revered entity in their society, a father figure who guided his children and led them along a proper path towards prosperity. Ur-Nammu's son Shulgi is considered the greatest Neo-Sumerian king. He not only continued his father's policies diligently but also pushed them further with a much stronger aptitude. He was ambitious to groom the policy domain of Sumerians. In order to impress his people, King Shulgi undertook a run of 160 kilometres between Nippur and the capital city of Ur and back again in one day so as to be able to officiate at the festivals in both the places, apparently, creating a sense of awe. Admiration was central to the governing power of the kings of Ur. There are many such important traits of Sumerian culture. The Sumerian language is considered the oldest linguistic record which persisted as a written language in cuneiform for almost 2000 years from about 4000 BCE to 2000 BCE. The *cuneiform language* is based on pictographic tablets of clay. They depicted their ideas by pictographs which included symbols that stood for words and sounds. Sharpened reeds were used to scratch the symbols on wet clay, which on drying became tablets. It was adapted into *Akkadian* language and dominated Mesopotamia for the next about 1000 years. This expanded outside Mesopotamia also for another 2000 years. The Sumer people used a script, yet they are credited to have developed a writing system and used it extensively for written communications. This system came to be known as cuneiform. According to Kramer, this script was borrowed by the subsequent civilisations and used across the Middle East for the next 2000 years. Writing is one of the most significant cultural achievements of Sumerians. Owing to their literary and linguistic excellence they were able to meticulously maintain the records from rulers to farmers and ranchers.

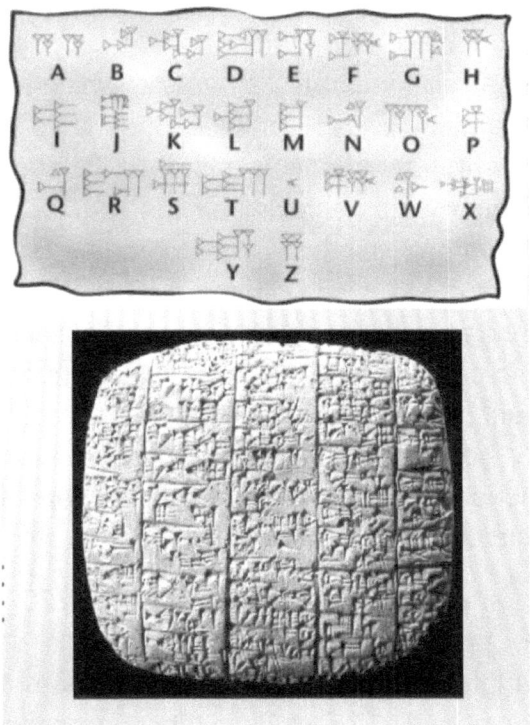

Cuneiform Writing on Clay Tablets – Ancient Sumer Civilisation

Sumerians were endowed with a rather uncommon flair for technological inventions. The Sumerians astutely recognised the limitations that were there in the region. They had a very limited number of trees, virtually no rocks or metals. They intelligently identified a substitute in clay and used it to make bricks, pottery and tablets for keeping their records. Sumerians were noted for building works which they undertook probably for extending the legacy of their predecessors the Ubaids, dating as far back as 5200 BCE. Later around 3400 BCE they started building religious structures on a grand scale and refined architecture with prominent features like flat roofs and arched doorways while the walls were built using roasted mud bricks mixed with marsh reeds. With their refined taste and adoption of finery in construction of houses, the Sumerians also built imposing brick columns, complicated mosaic structures, mural paintings, terracotta ornaments and bronze accents. The temples were specially decorated by artistry which had a significant presence of naturalism. Except for water, they were short of the rest of naturally occurring construction materials. There were no rocks in the region which were replaced by the roasted mud and reed bricks. The Sumerians used the roasted/dried clay bricks made by using clay and reeds in pre-made-up moulds. The houses and temples constructed with these bricks were not very durable but they were able to construct the buildings in large numbers leading to the establishment of larger cities. They were also able to effectively substitute other construction materials like stone and timber that were conspicuously absent in the region or were too little. They substituted stone carvings with metal casts.

They built temples of fixed norms around 2200 BCE, largely resembling the pyramids. These temples were stepped, either square or rectangular in shape, had an inner chamber, the sloping sides and stood about 170 feet high. The Sumerians called them *Ziggurats* which were also prominently decorated with terrace gardens. The hanging garden of Babylon was one such imposing piece of creation which showcases the level of sophistication in their life system. They did not lack grandiosity. A palace with 200 rooms in the Sumerian city of Mari is an example of their ambition to achieve the best in life.

The Sumer people had the knowledge of anatomy and probably practised surgery for treating the sufferers. They also practised their own system of medicine which was based on magic and herbalism. The Sumerians were wise to implement the inventions developed by other groups and societies in other places. Such inventions were implemented by them on a larger scale. Their strategy was to handle goods in large volumes and trade them with other people. They had incredible organisational capacity, an attribute that ensured their success in almost every arena of life. There was something in the Sumerian identity that drove them to dream big and think ingeniously. The Sumerians, as further explained by Kramer, "Spiritually and psychologically, they laid great stress on ambition and success, pre-eminence and prestige, honour and recognition." Their approach towards life was constructive, a trait which they passed on to their future generations. The Sumerians created yet another *tour de force*—the modern practice of checking one's horoscope was fashioned by them. It is believed that the astrology and the Zodiac signs in some form were conceived by Sumerians in their flourishing period but no authenticated theory of astrology can possibly be attributed to their credit.

They developed the turning wheel, a device which allowed them to produce pottery on a large scale while others made them by hand. That enabled them to produce a large number of items such as containers for worker's ration, also used as the granaries. This represents the marked headway in the field of planning, conservation and optimal use of resources.

The Sumerians are dotted as adopting very productive farming, particularly with advanced irrigation. They designed and implemented a complex system of canals, constructed dams using reeds, Palm trunks and mud or wet clay. The dams even had gates that could be opened and closed to regulate the flow of water. They developed a kind of plough for farming with far better results. They even wrote instruction manuals on *Ninkilim,* the deity of arable land, wild life in general and vermin in particular. This incorporated the details on how to use various types of ploughs. The Sumerians had worshipped the Goddess of field rodents, in order to protect the grains from being eaten. They composed a prayer also and recited it to pray to the Goddess. Although the Sumerians are not credited with inventing the wheeled vehicles, they probably developed a two-wheeled chariot pulled by a team of animals who were reined by a driver for moving it. According to another school of historians, such vehicles were probably used for military purposes or for ceremonies rather than as means of movement of citizens in the countryside.

The Sumerians perceived the significance of volume in the business model. The other cultures in the Middle East also prepared textile items by weaving the fabric in smaller numbers. On the contrary, the Sumerians opted to do it on a much larger scale. Sumerians virtually laid the foundation of modern manufacturing companies. They were the first to cross kin lines and form large working outfits for producing larger volumes of textiles, a model that has come to stay today in modern manufacturing companies. Their language and writing were the main supporting tools in the development of trade links and the management of their force of agents and representatives. The accuracy in communication with their own people stationed far and wide must have been very effective, leading to worthy results.

The Sumerians were the first people who used copper. They made spearheads, chisels and razors, etc. using copper. The artefacts and utensils were also created in copper, which included items like panels depicting fanatical animals such as an eagle with a lion's head. As stated by Kramer in his work, the Sumerians were knowledgeable metallurgists. They developed furnaces and heated them by reeds wherein the temperature was controlled by a device with an airbag that emits a stream of air, the bellows, similar to what we see today with small smelters.

As described before also, the Sumerians developed a formal numbering system based on units of 60, as elaborated by Robert E. and Carolyn Krebs in their highly acclaimed book, *"Groundbreaking Scientific Experiments, Inventions and Discoveries of the Ancient World"*. At first, they used reeds to keep track of the units; later on with cuneiform in place, they used vertical marks on the wet clay tablets. They invented a *sexigesimal system* for depicting time, a system based on the number 60—a minute comprising 60 seconds and an hour comprising 60 minutes. Postulated as the basis of civilised human society, they also divided day and night into spans of 12 hours each. They also set limits on a 'workday' or hours of working in a day which is now called a 'man day'. The concept of holidays or the 'days off' was also established. This truly laid the foundation of mathematical calculations for future civilisations. Trade by Sumerians flourished with communities far away. The Sumer inhabitants are thought to be one of the potential targets for extensive trading carried out by Indus-Saraswati people. The Sumerians' innovations gradually spread and in due course of time led to the development of the modern technologically advanced world that we live in today.

It is rather hard to accept that a civilisation as highly advanced as Sumer, suddenly disappeared rather inconspicuously. There are many opinions, which assign different causes responsible for their extermination. Following the Ur III period and the fall of Ur, many Sumerians migrated north. Sumerian was no longer spoken as a language, though it was still written, having been largely replaced by the *Semitic Akkadian*, marking the virtual end of the Sumerian culture. Their legacy, however, continues in many aspects of civilisation which modern-day human society takes for granted as always existing.

American Archaeologist and Anthropologist Robert McCormick Adams must have visualised the disaster when he describes the site of an ancient Sumerian Civilisation located on the central flood plain of the Euphrates River, a desolate and disgruntled area now outside the frontiers of cultivation as, "tangled dunes, long disused canal levees, and the rubble-strewn mounds of former settlement contribute to only low, featureless relief. Vegetation is sparse, and in many areas it is almost wholly absent .... Yet at one time, here lay the core, the heartland, the oldest urban, literate civilisation in the world." It is generally propounded that the early Sumerian Civilisation of the fourth millennium BCE had advanced far beyond any of the previous ones. Their irrigation system, based on sophisticated engineering concepts, created highly productive agriculture that helped in the production of surplus food which subsequently supported the formation of cities. Sumerians had the first cities and the first written language, the cuneiform script. They represented an extra-ordinary civilisation but failed to cure an environmental flaw in their irrigation system. Water from dams and storage reservoirs was diverted onto the land to raise abundant crop yields. Some water was used by the crops for their growth, some got evaporated naturally and some percolated into the groundwater-table. The water-table must have been already higher, replenished by the surrounding river system of powerful rivers like Tigris and Euphrates. Over time this percolation raised the water-table to reach a few feet below the surface level thus restricting the growth of deep-rooted crops. Later on the groundwater-table, when fed more, climbed to a few inches from the surface. At one stage water submerged the soil and humus and began to evaporate into the atmosphere, leaving the salt in the water behind. Further on, the accumulation of salt hampered the fertility of the soil, consequently resulting in flagrantly poor crop yield. The environmental flaw was the absence of any arrangements for draining the water from the groundwater-table. The initial response to declining wheat yield was to shift to barley, a more salt-tolerant crop. Eventually, the yield of barley also declined. The food supply shrank to the brim, undermining the economic foundation of a great civilisation, leading finally to its complete eradication. The complete untimely annihilation of this rich civilisation is a great chronicle to visualise what environmental disaster can do to the very life. This type of environmental aberration was probably missed by the Sumer population. It was the excess of water that played the deciding role, leading to the shortage of food supplies and subsequently the collapse of Sumer Civilisation.

## Chapter 8

## Calamitous Occurrences with the Mysterious Mayans in Outer World

*Mesoamerica* is the historic region that comprises the modern-day countries of Nicaragua, Costa Rica, El Salvador, Guatemala, Belize and central-to-southern México. This region has remained populated by various groups such as *Olmec*, *Zapotec*, *Toltec*, *Aztec* and Maya people. The prominent traits that signify the cultural status of the region include the domestication of maize, beans, spices like vanilla and fruits like Avocado and more or less similar architectural style among them. Mayans brought in much greater impetus to the cultural evolution in human society and have become a subject matter of continued study by the modern archaeologists, Palaeontologists in addition to social scientists.

Centuries before the Europeans had settled down in the region extending from southern México through Central America, often called *Mesoamerica* hosted an advanced civilisation that flourished in almost all walks of human life. This was the mysterious Maya Civilisation. The Central American region, in fact, was occupied by hunter-gatherers from about 1800 BCE to 250 CE, a period which led to the creation of village life which in turn gave rise to early Mayan cities. The land area where the Mayans flourished lay between the Gulf of México on the west, the Caribbean Sea on the east and the Pacific Ocean on the south. The period which signifies the peak of Maya Civilisation is called the Classic period, ranging from 250 CE to 900 CE. The Maya Civilisation had spread over an almost continuous territory of about 311,000 square kilometres. The entire territory can be broadly divided into different areas, the Septentrional area which is mostly the area with scrub vegetation and no or little water, the only source being the Lake Chechancanab. This area encompassed the Yucatan peninsula and included the sites like Uxmal, Labna and Chechen Itzá, etc. After the lowland city-states collapsed, ending the Classic period, Yucatan peninsula became the venue for Maya Civilisation to continue and thrive. This area lay on the north of the territory. Adjoining the Septentrional area lay the central area which included north-western Honduras, through Peten region of Guatemala and into Belize and Chiapas. This became the heart of the Classic Maya Civilisation. The cities included in the central area are Copan, Yaxchilan, Tikal and Palenque. The third, as already stated, is the area of Guatemala highlands lying on the southern end of the territory stretching up to the coast of the Pacific Ocean. This part signifies the early Classic period which mainly included Guatemala.

Maya is a civilisation enshrouded in mysteries and their significant population still inhabits, as the civilisation did not vanish completely. While some other population in *Mesoamerica* formed a scattered pattern of dwelling establishments, the Mayans created densely populated villages. The Maya developed and lived mainly in Yucatan peninsula of México, and what is now Guatemala; Belize, parts of Tabasco and Chiapas states of present México and western part of Honduras and El Salvador. Like their villages, the city-states of Mayans were also densely populated. In that expanse of land, the Maya lived in three separate sub-areas, elaborated before; and the inhabitants of three regions displayed distinct environmental and cultural differences. The Maya are a congregation of an indigenous group of people living in Central Americas. They are not a single entity or a single community or a single ethnic group. They speak many languages such as Yucatec, Quiche, Kekchi and Mopan, considered to be Mayan languages, in addition to Spanish and English. However, they are an indigenous group closely attached to their past spanning over 2500 years as well as to the events of last about five centuries.

The earliest settlements of Maya people came up around 1800 BCE. This is known as formative or preclassic period. From the initial period, they resorted to agriculture. They cultivated crops like maize, beans, squash and manioc or the cassava. As the population increased measures of enhancing the food supply were also put in place. Mayans started truly effective farming in preclassic period around 1000 BCE. This period saw the cultivation of more productive forms of maize, the main source of carbohydrate. A process named "nixtamal" which comprises soaking of maize in lime water or something similar to it before cooking immensely enhancing its nutritional value, had become a recurrent practice among Maya clans.

It is noticed and almost established that the results of a meteorite striking the earth were staggering, particularly on the land encompassed in Yucatan peninsula of México. The cracks developed in the limestone terrain gave rise to several *cenotes* in due course of time. These water pots of enormous capacity played an extremely significant part in the life of Maya people. They provided water for irrigation and super clean water for drinking. In the decade of seventies, the researchers noticed the presence of the remains of elaborate irrigation canals in the wetland areas. The Maya people built canals to bring water from wetland areas to create new farmlands. In other places where the stones and rocks were quarried for the construction of their massive pyramids and temples, the resulting pits, the disused quarries had rainwater accumulated in them, channelled into the canals to reach the farmlands, used extensively for irrigation. This civilisation was fully aware of the value of water and was managing it methodically. The centre at the Mayan city Palenque had what was arguably the most unique and intricate system of management of water anywhere in the contemporary period. There were places in Mayan territory with no such water source present. The Mayans in such places turned to the storage of water for domestic use, by using man-made cisterns. These cisterns were similar to today's large earthen pitchers shaped like bottles, called '*chultans*'. These cisterns were lined with lime plaster to prevent any seepage. The storage was done with certain restrictions, and distribution would be accompanied by some kind of rationing on the use and supply of water. Any hiatus in the process of filling the cisterns or wastage of water or its excess use by the residents, would lead to a disastrous outcome; the Maya societies were fully aware of such factors.

All their efforts to lead a contented and comfortable life were focussed on the judicious use of water, which underlines their concern over its limited availability. The cultural practices by Maya people began expanding in highland and lowland regions, during the middle preclassic period which lasted until 300 BCE. This period also saw the rise of *Olmec*s, a major *Mesoamerican* civilisation, more or less contemporary to Mayans while some consider *Olmec*s clearly younger to Maya. Maya lasted the longest and continue even on this day. *Olmec*s were followed by Inca in modern-day Peru who in turn were followed by *Aztec*s in modern-day México. Maya accompanied them all and beyond and are considered to be the greatest Mesoamerican civilisation. Of many traits the Mayans learnt and obtained from *Olmec*, included were the number system, the worship practices and the creation of their famous calendar. The Classic period began around 250 CE and lasted to about 900 CE which is regarded as the golden age of Maya Civilisation. The growth during this period was exponential. They built great stone cities and monuments during this period. This civilisation reached its peak around 600 CE. They built as many as 40 cities during the Classic period, including Tikal, Bonampak, Calakmul, Coba, Chunchucmil, Chichén Itzá, Szibilchaltun, Chiapas, Palenque, Oxkintok, Dzibilchaltun Yucatan, Tulum, Uxmal Yucatan, Axchiltan Chiapas in México, El baut Ceibal aka Seibal, Dos Pilas, Iximche, Ixkun, Kaminaljuyu, El Mirador Piedras Negas, El Peru, Mayapán, Naachtun, Nakbe, Nafanjo, Q'markal, San Bartolo, Yaxha in Guatemala, Caracot in Belize, Copan in Honduras, and Tazumal in Chalchuapa region in El Salvador. Almost all these cities housed about 5000 to 50000 people. The total population of Mayans had ranged from about 2.5 to 3 million people during the peak period. About 6 million Maya people live today and are the largest single block of indigenous people.

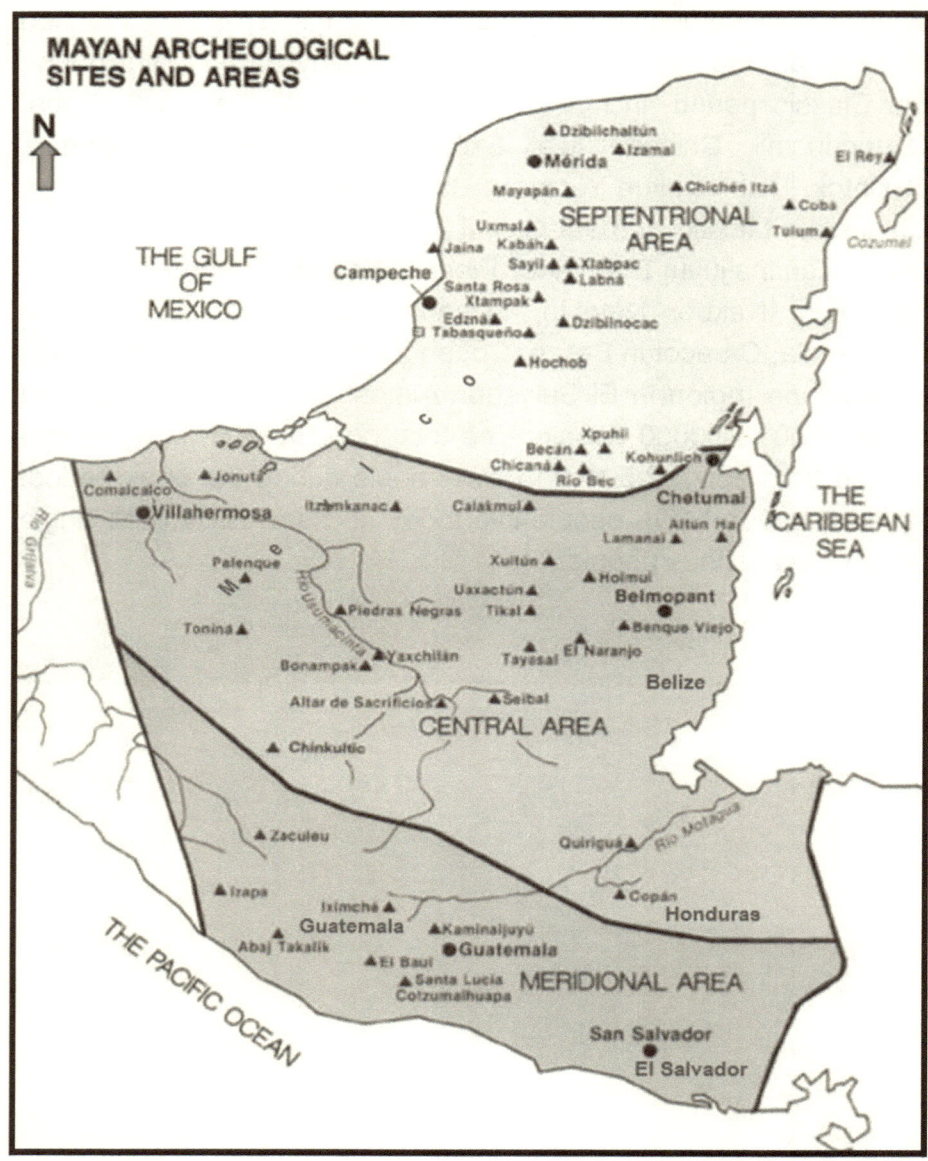

Tentative Location of Inhabitations Developed by the Mayan Civilisation: Modern-Day Guatemala, El Salvador, Honduras, Belize, Parts of México

There are about 3000,000 Mayans who live on tropical Yucatan peninsula, called Yucatecs, 120,000 in Tzotzil and 80,000 in Tzeltal in the highlands of Chiapas in México. In addition, places like Chontal and Chol in México are also inhabited by Mayans. In Guatemala, some large groups of Mayans inhabit places like Quiche and Cakchiquel. Belize also has Maya citizens, known as Kekchi Maya. In all, there are about 30 Mayan people groups throughout Central America speaking different and virtually completely unintelligible languages of Maya family. American archaeologist, anthropologist, epigrapher and author, Michael Douglas Coe, an acclaimed authority on Maya, Professor in Yale University in his book "*The Maya*" published by Thames and Hudson in 2011 revealed various details of Mayan culture. Distinctive traits of Mayan people characterise their culture prominently. The published civilisation enjoyed the optimal growth in the Classic period. The Maya achieved the intellectual, artistic and cultural heights, unrivalled by any other in the new world, just a few could match in Europe. Coe, states "Large populations, a flourishing economy, and widespread trade were typical of the Classic Mayans". Warfare was also quite common. The earliest Maya people were farmers and the economy was based on farm output. As already, stated in order to cater to their basic requirements, the Maya people cultivated maize, beans, squash and cassava or manioc. The middle preclassic period continued until about 300 CE. This period witnessed the beginning of the expansion of their territory, undertaken by erstwhile Maya population.

Simultaneously the *Olmec*s started to stabilise in present-day Mexican states of Veracruz and Tabasco while people of other personae such as *Totonac*, *Aztec*, *Zapotec* and *Teotihuacan* also existed in their respective slots. The Mayans in this period undertook agriculture, quality construction of pyramids built using a grid system decorated with architectural finery and inscribing of stone monuments. This amounted to a sure sign of urban planning. City of Mirador in Peten region of the Republic of Guatemala was

Ruins of a Pyramid Built by Ancient Maya Civilisation: Location EL Tazumal, Chalchuapa, El Salvador

considered to be one of the greatest cities ever built in the period. It was larger than the Maya capital Tikal of the Classic Maya period. They also built plazas, palaces, temples and pyramids. They built courts called Pitz for the famous Maya ball game Ulama. This game was in fact, played in a court shaped 'I' with a rubber ball ranging in size from that of a softball to a football. The court itself represented Maya cities and symbolised the city's wealth, power and influence. In fact, the ball game was a part of civilisation's political, religious and social life. The ball game also gave neighbouring cities a platform for settling the disputes and were considered an alternative to war. Based on a mythological folklore they believed that it was necessary to play the ball game for their own survival. It provided an opportunity to show devoutness to the gods by sacrificing captured kings and the high ranking personnel or the losing opponents in the game. The ball game formed an imperative part of Mayans' life.

The Maya people had a deep trust in worshipping various gods. Most of their religious bend was directed to nature. However, they mainly worshipped God Sun, called *Kuhul Ajaw*. The Mayan Moon Goddess was named *Ix Chel*; the Rain God as *Chaac*; the Mayan God of Flora as *Yumil Kaxob*; the Mayan God of Forests and Maize as *Yum Kaax*; the Yucatec Mayan God of Death as *Yum Cimil*; and the ruler of the heavens was known in their vocabulary as *Itzamna*. Maya Civilisation worshipped and constructed a pyramid for their most powerful God, *Kukulkan*, the snake God whose name means "feathered serpent". God *Kukulkan* was widely revered by Mayans in the latter part of the Civilisation. The Maya societies were headed by kings or *"Kuhul Ajaw"*, which meant holy lords, who claimed to be related to gods and followed a hereditary succession. They performed the elaborate religious ceremonies and rituals which were important to Maya culture. In fact, they were thought to serve as the mediator between the gods and people on earth. In order to serve and please the gods, the Maya even resorted to human sacrifice also. It will not be out of context to state that Maya people are believed to have desired to capture the strength of their gods, to become divine and garner more and more power to be able to obtain the remedies for human problems. At the peak of their civilisation, the Classic Maya built many temples and palaces in a stepped pyramid shape. Their buildings were also decorated with inscriptions and incorporated conspicuous relief. Endowed with construction expertise and architectural excellence of their stepped pyramids, the Maya people were regarded as great artists of Mesoamerican world. They made significant advances in astronomy and mathematics. They aptly used zero already invented by then. The Mayans developed a complex calendar system like calendar count covering 356 days. Later on, they created the Long Count Calendar designed to last over 5000 years.

The ancestor of three Maya languages in use presently had developed a complicated calendar with an 18-20-day month plus a set of five days. The calendar system also included what is called a "long count". This kept a track of time by using different units that range in length from a single day to a long period using different units. In total contrast to a largely talked-about belief that the Mayans in their calendar predicted the doom of the world in 2012, they probably expressed the alleged prediction in some other units like millions of years. The other popular belief of Maya Civilisation was to have acquired the ability of vanishing by themselves also has gone haywire. Many of their urban establishments were abandoned about 1100 years ago but others developed, instead. Chichén Itzá for example developed only after that.

Their writing system, the hieroglyph system could be deciphered in the early-to-mid-20th century. The remains of their buildings, architecture and art including the stone inscriptions and carvings have been the sole source of understanding the true and unadulterated Maya Civilisation. They wrote books known as 'codices' using the paper made from tree barks. Four such codices are known to have survived. Some pieces of evidence of the earliest use of rubber and chocolate have also been traced. Probably the most intriguing fact about the Maya was their ability and efforts to build such an advanced civilisation in tropical rain forest climate. *Teotihuacan* people in highlands were contemporaneous to Maya in their peak period. Surprisingly, Maya chose dry climates where centralised management of water resources was adopted. In southern Maya lowlands, there were a few navigable rivers facilitating the trade and transport, also serving the requirement for an irrigation system. The lowlands with diverse environment proved disappointing for foreign invaders for not being able to get precious metals like gold and silver. Instead, this became advantageous for Mayans. They exploited the region's natural resources, such as limestone for construction and obsidian, a volcanic rock for making tools. They exploited Jade in the region and quetzal feathers for decorating elaborate costumes for society's high rankers and nobilities. The marine shells collected were used as trumpets in warfare and religious ceremonies.

A gradual collapse of the Maya Civilisation presumably began somewhere around 1100 CE with the downfall of northern Maya cities. This followed the end of the classical phase of Maya history. The collapse of Maya took a few hundred years exhibiting their undisputed cultural strength and is now a subject matter of study. There are many reasons assigned to settle the problem. One is overpopulation causing contest leading to conflicts for available food supplies. The food grain production remains virtually the same as before but there might have been a conspicuous increase in the consumers thus defying the principle of sustainability. Warfare, shifting trade routes, extended drought and environmental degradation were some other possible causes. This Mesoamerican civilisation collapsed and perished owing to some reason most likely attributable to some natural calamity vis-à-vis environmental disaster. It is likely that a complex combination of factors was behind the collapse. Overpopulation is a factor that might have destabilised the vital balance of life. Although there are many hypotheses ascribed to the progressive collapse of this exceedingly prosperous and remarkably well-informed civilisation presumably with outstanding knowledge of astronomy and warfare also. Raging war could be one causative factor. The Mayans fought with the other contemporary civilisations, consequently conceding the defeat, left the venue, while another school of historians propound that the ruling establishment faced stiff and bloody opposition from working or labour section of the society and could not sustain. Some neo-enterprising section of historians even went to the extent of imagining an alien attack on Mayans. However, the most convincing hypothesis is the falling food availability which is the major cause attributed to the untimely demise of the civilisation. The food scarcity may have triggered the conflict among various Mayan cities as they must have competed for food. Mayans, the proven experts of agricultural farming resorted to indiscriminate cutting and burning of trees and plants thus removing the vegetation canopy on large areas. Apparently, it was deforestation and the erosion of soil that undermined the agriculture. This exposed the earth surface to direct sun rays which drastically reduced the earth's capacity to return the water vapours to the atmosphere. Subsequently, there must have been sustained drought which adversely affected the food availability and the majority of Mayan population probably starved to death. This has now the staunch support of scientists who base their interpretations on the analysis conducted on minerals collected from great "blue hole", a giant submarine sink-hole off the coast of Belize, an important Maya area and the surrounding lagoons. This research has now established that from about 800 CE to 900 CE there was a horrifying drought lasting for about a century that engulfed huge areas inhabited by the erstwhile Mayans. This kind of situation appeared intermittently, in waves.

The hypothesis favouring drought had some experiments which not only support it but also prove it beyond doubt. Gratefully, the experiment conducted by Geologist Dr. Mark Brenner of University of Florida, USA yielded groundbreaking results to conclude 'in one simple statement' that it was a horrifying drought occurring intermittently in the majority of Maya territory that brought off their collapse. The Puuc on Yucatan peninsula in present-day México, had no water of its own. This was the venue where the Mayan Civilisation thrived. On the edge of the region there is a lake called *Chichancanab*, also known as Laguna *Chichancanab*. It is located in the northwest of the state of Quintana Roo. The lake is endorheic and water in it is saline. The salinity is mainly due to the presence of mineral Gypsum, the calcium sulphate. It is a cenote which has surface connections to subterranean water bodies. The clay bed at the bottom seals the water off the bedrock. The level of surface water in the lake is entirely determined by the balance between rainfall and evaporation. The core samples obtained from precision core drilling, done at the bottom of the lake were studied by Dr. Mark. Brenner. Samples from lakes are wonderful recorders of past environmental history. The core retrieved shows the depositional chronology. In the core samples of floor deposits, in Lake *Chichancanab* down to a depth of few metres, the brown mud layers were found interbedded with Gypsum layers. These Gypsum layers must have been deposited on the evaporation of water in the lake, indicating the situation of a drought. Gypsum layers are seen intermittently in the core of mud retrieved from the bottom of Lake *Chichancanab*. Gypsum from various layers in the core samples of the Lake was subjected to carbon dating to assign the period of its formation and deposition. The age coincided with the fall of the Maya Civilisation. In fact, the Gypsum in the bottom of Lake *Chichancanab* was truly in situ as the lake is endorheic which means that the Gypsum had not come there from any other source, outside the premises of the lake. In addition to Gypsum, there are snail shells also preserved in the mud, retrieved with the core. These shells reveal the environmental history with greater accuracy. Inside aragonite which constituted the snail shell, two distinct *oxygen isotopes* are locked. The determination of oxygen isotope ratio was used to determine the situation—dry or wet. In the times of drought, one type of isotope dominates and can be detected in the chemical composition of the shell. The analysis for finding out the number of isotopes revealed that the type which indicated the drought was far more than usual times when the lake had water. A series of drought-like situation was detectable and they were of enormous intensity. The snail shells also exhibited that the things were incredibly dry. The study of lake sediments provided unambiguous evidence for severe drought that persisted for nearly 200 years from $8^{th}$ to $10^{th}$ century that is about 800 CE to 1000 CE. This was probably the driest period of middle-to-late Holocene epoch. It is thought that the drought of such a magnitude had not occurred in the last about 7000 years. The period coincided with the extermination of classic Maya Civilisation. Yucatan peninsula region was a mute witness to such a grisly disaster.

The Mayans used to store water which according to research could sustain for three months during the period of drought and with stringent rationing of water, maybe for a few months more. The repeated occurrence of Gypsum bands at frequent intervals marked the incidence of drought in the region, time and again. What would have happened to the Maya inhabitants on the peninsula is easy to comprehend. By the time the population would be able to recover from one incidence of 10 or 20 years, they were pounded by another bout. Ultimately they had no option left but to leave the place and their well-groomed legacy. It may be concluded that the incidence of repeated drought for longer durations brought about the collapse of such a prosperous and knowledgeable civilisation. The situation of drought prevailing intermittently for long, emanating from environmental degradation leading to climate change, left the population, facing severe starvation, resulting in a hungry civilisation which made immense contribution to the development and improvement of life standard of humanity, became exterminated, though not entirely.

Chapter 9

# The Appalling Events in Rapa Nui, an Exemplary Case of Gradual Despair and Sustainability

An enigmatic island located at the southeasternmost tip of the Polynesian Triangle in the South Pacific Ocean, geologically one of the youngest inhabited territories on the earth, is known as Rapa Nui, the Easter Island. Sometimes also called Paaseiland, it is a volcanic island, rather small, admeasuring about 163.6 square kilometres—23 kilometres in length and 11 kilometres in width. Easter Island is one of the most remotely inhabited islands in the world. The island sits atop the 'Sala y Gomez' submarine ridge, which trends eastwards from the east Pacific rise. In 1888 it was taken over by the Republic of Chile. In fact, the British regime recommended Chile to claim it to prevent France from doing it first. The inhabitants of the island were granted Chilean citizenship in 1966. It is remote to the extent that the nearest island Pitcairn Island is located about 2075 kilometres, inhabited by just about 75 residents; another island Mangareva is about 2606 kilometres away. Easter Island is about 2500 Kilometres away from the Chilean capital of Santiago, a five-hour journey from there to Mataveri airport, at Hanga Roa on the island, it's lone. It gained the constitutional status of "special territory" in the year 2007. The island was bestowed upon the *locus standee* of the World Heritage Site by UNESCO, with the majority of the area on the island protected as Rapa Nui National Park. The nearest continental point to Easter Island is a part of Region Valparaiso and state Ista de Pascua, in central Chile, located about 3510 kilometres away. The island is composed of three major vol*canoes* and more than 72 subsidiary vents. *Rano Kau* , and *Poike volcanoes* are located on the southwest and east tips of the island respectively and are dated to Pleistocene age. *Rano Kau* volcano opens up on the surface as a one-kilometre-wide crater. The third and youngest named *Teravaka,* which also marks the highest point of Pliocene to Pleistocene age, is located at the northern leg of the triangle, the shape of the island. Prominent vents are also named as Rano Aroi and Kaitiki, etc. located in the north and east portion of the island. Rano Rakaru volcano hosted the rock quarry. The last stage of volcanic eruptions came up from multiple rift zones oriented along the axes of the island. The lava flow that occurred about 2000 years ago is the last proven volcanic activity which took place at Hiva-Hiva near the west-central coast.

The rocks found on Easter Island are invariably extrusive igneous rocks. The lava spewed out by three vol*canoes* formed *basalt*, *Trachyte*, *volcanic tuff* and *Scoria*. The voids in *basalt* flows have also created the ideal situation for crystallisation of cryptocrystalline silica-mineral quartz in various forms. The topography signifies the rolling hills, three volcanic craters and numerous vents. The lava flows, domes, cinders and pyroclastic cones are present. In fact, the rocks represent the entire compositional range from *basalt* to peralkaline *Rhyolites*. The significant characteristic of the magma is two to three orders of magnitude lower in viscosity which is due to high alkali and halogen content in it. The soil on Easter Island contains *bacterium Streptomyces hygroscopicus*. A derivative of this content is known as *sirolimus*. Pharmacologically it is named *Rapamycin*, named after the island's Polynesian name. This pharmacological product is used to prevent the rejection of organ transplant. Additionally, it is also known to have an influence on ageing in the lower organisms by also disrupting the influence of an enzyme known as TOR. This has been experimented on the mice where it has been noticed that *Rapamycin* feeding could extend the lifespan even when the medication is started late. More intensive research work augmented might lead to much sought after anti-ageing or life-enhancing pharmaceutical products. This path-breaking research is conducted by a dedicated team of 14 researchers, led by David Harrison working in Jackson Laboratory at Bar Harbour, Maine, Sacramento County, California.

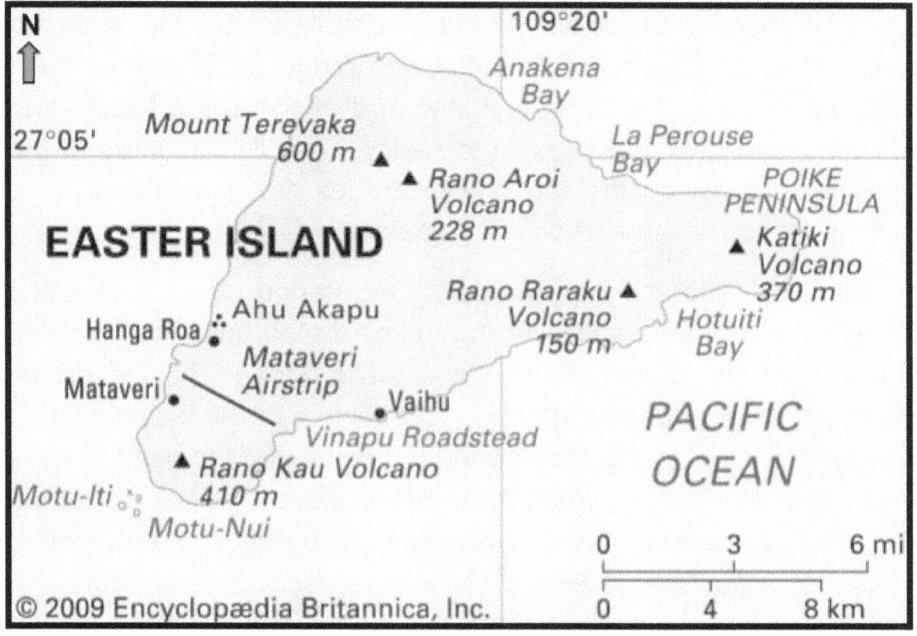

Tentative Location Map of Easter Island, the Rapa Nui or Isla de Pascua: Now a Part of Republic of Chile

The phrase "*Te pito o te henua*" is said to be the original name of the island which when translated by French ethnologist Alponse Pinart means "the navel of the world". Spanish people gave the island another name 'Isla de Pascua'. Later on, the aboriginal Polynesians, the sailors from Tahiti, presumably the first human inhabitants gave the island its Polynesian name, "Rapa Nui". However, some believe that the people from Marquesas Island in French Polynesia were the first settlers on Easter Island. Yet another suggestion is that a second migration may have originated from Inca-dominated regions in South America. In 1950s, legendary Thor Heyerdahl had stirred the scientific fraternity by propounding that the Easter Island had South Americans as its first settlers. This theory was not substantiated by many scientists. Cultivation of sweet potatoes or yams by Rapa Nui inhabitants is neither endemic to Rapa Nui nor to Polynesia but is very common to South America. The clay tablets, inscribed with Easter Island script, known as *rongo-rongo*, used by the inhabitants, found on the island was also found around *Lake Titicaca,* a sacred venue for *Inca Civilisation* in the Andes range. With various traits strongly exhibited by the inhabitants of the island, the Polynesian ancestry appears to be closest to reality. The DNA analysis of the skeletal remains on the island also proves that inhabitants were strongly linked to modern-day Polynesians.

It is an island, with incredibly mysterious rock monoliths chiselled as human heads, having torsos rooted below the ground. These stone sculptures are known as Mo'ai *(Moai)*. Thor Heyerdahl identified three separate epochs in the history of the island—Early, Middle and Late. In the early period the carving of the giant statues was not done. They devoted their skill to erect altar-like platforms, the *'Ahu'* comprising large stones shaped by cutting and joining, constructed facing the ocean. They were astronomically oriented, and they aligned these rock-cut daises with the alignment of the sun. The middle period which lasted from about 1100 CE to 1681 CE was devoted to quarrying, carving and transporting them to various venues across the entire island and placing them on their respective *Ahus.* The late period saw the sudden end to all carving activities in the quarries under *Rano Raraku* volcanic crater by consolidated tuff on the slope of mount Maunga *Terevaka* or mount Terevaka. One school of scientists put the pieces together and propounded a hypothesis that island society had undergone a period of rebellion, civil war and famine—maybe all of them together or none of them at all. But something undesired was certainly around resulting in complete halt of the inhabitant's favourite activity.

The first human inhabitants on the island, the Polynesians arrived on the island between 700 CE to 800 CE or even before. They brought with them bananas; taro or kalo a kind of corn, the sugarcane, a tropical plant from Hawaii and paper mulberry as well as chickens and Polynesian rats. After settling down on the island they included fish also in their diet, particularly the Tuna and *Dolphins*. Herding to this lonely and remote island by them was too enigmatic to fathom, until it was realised by social scientists that high level of *Ciguatera* fish poisoning became rampant across the region. Initially, nobody knew what caused them such misery. Ciguatera possesses some poisonous material in its flesh that causes diarrhoea, itching and other skin ailments, sometimes leading to death. The sailors might have tried to duck this misery and headed towards the island.

The first European visitor, the Dutch explorer Jacob Roggerveen, reached the Easter Island on Sunday, April 5, 1722. Since it was Easter Sunday, Roggerveen had all the reasons to name the island as Easter Island. After about 48 years of Roggerveen's pioneering voyage to Rapa Nui, the Spanish Navigator and cartographer *Felipe Gonzalez de Ahedo* visited the Easter Island in 1770. He claimed the island for Spain on a document written by islanders on *rongo-rongo*, a system of glyphs, undecipherable Rapa Nui script. British explorer, navigator, cartographer and a naval Captain *James Cook* and French naval officer *Jean-François de Galaup, Comte de Lapérouse* visited the island in 1774 and 1786 respectively.

The Easter Islanders spoke Spanish as well as traditional Rapa Nui language. It is also known as *Pascuan*, currently written in Latin script and is classified as an Eastern Polynesian language. However, majority of the people speak Spanish which pushes the Rapa Nui language in the minority slot. The Rapa Nui language appears to be leaning and moving more heavily towards Spanish. Rapa Nui language's version, *rongo-rongo*, discovered about 200 years ago is believed to be its older version. However, it is not yet clear whether *rongo-rongo* is a form of writing or some different form of cultural manifestation.

The most redeeming feature of Rapa Nui is the large monolith stone sculptures created and located at various places all over the island. They were given the name Mo'ai *(Moai)*. Carving out the stone statues was an integrated part of Polynesian religious virtues. It was an accepted belief that the Polynesians were the first to inhabit the Easter Island, nursing the Polynesian culture and like them sculpting the Mo'ai *(Moai)* was also indigenous to Rapa Nui. In contrast to this popular and almost a proven idea, Thor Heyerdahl, thought that inhabitation in the island had been the effect of Incan migratory wave. The *Inca Civilisation* inhabitants on the island, he propounded, introduced the style of sculpting the stone statutes similar to the one extensively adopted back home. As already stated this did not catch up with most of the scientists. The inhabitants of island Rapa Nui built as many as 887 human figures sculptured from different rocks. They were not only created but also transported to various places all through the span of the island. The islanders strongly believed that a symbiotic relationship exists between them and their dead. In their relationship, the dead provided everything that the living needed, such as health, fertility of land, animals and fortune, etc. Reciprocating, the 'living' provided the dead with a better place in the spirit world. Most settlements were located on the coast and mostly the Mo'ai *(Moai)* were erected along the coastline. Their specific position is probably very significant. They were so placed as if watching over their descendants on the island, taking care of the respective family, with their backs towards the spirit world presumably to ward off the evils. Every family probably built its own Mo'ai *(Moai)*. The islanders remained carving *Mo'ai (Moai)* for almost 300 years from about 1200 CE to 1500 CE.

Out of a total of 887 numbers, 834 were carved out from rock tuff, the consolidated form of volcanic ash, 17 were created from rock *Scoria*, 22 from *Trachyte* and 18 from *basalt*; all these rocks formed from the lava flows. Still unfinished which would have been the largest, named *'El Gigante'* may weigh up to 182 metric tonnes, equals to the weight of almost two Boeing jets with passengers on board and about 20 metres tall had it stood up at all. In the year 1868, they got toppled. The Mo'ai *(Moai)* are seen resting on large stone platforms called *Ahu*. The largest of them, in place, is *Ahu Vinapu*. Some Mo'ai *(Moai)* adorn red volcanic stone headgear called *Pukao*. The islanders probably performed this remarkable feat of sculpting the Mo'ai *(Moai),* during the year 1250 CE to 1500 CE. The main quarry was worked at *Rano Rakaru* hillock whose slopes still have quite a few numbers of Mo'ai *(Moai),* while hundreds were transported from there and placed on *Ahu*, the stone platforms, around the perimeter of the island. The Mo'ai *(Moai)* are mainly the living faces of the ancestors called *aringa ora,* while the deified ancestors are known as *aringa ora ata tepuna.* The statutes gazed inland across the land of the respective clan before they fell, probably due to erosional changes in soil profile around them. One school of archaeologists assign causes like internecine tribal conflicts or the European interference. These factors are held responsible for their prostrate positions which probably happened in the late 18th and early 19th centuries. Their creation and transport to various places on the island must have been an awe-inspiring graft and physical feat. The tallest finished *Mo'ai (Moai)* called *Paro,* was about 10 metres tall and weighed 82 tonnes. The shorter and squatter *Mo'ai (Moai), Ahu Tongariki* weighed about 86 tonnes. *'El Gigante'*, the heaviest and largest that remained unfinished lying in the quarry clears the sequence adopted in sculpting such monolithic classics. *'El Gigante'* the largest of the *Mo'ai (Moai),* was abandoned sometime around the year 1680. The average height of these massive stone works is 4 metres and the average weight is around 12.5 tonnes. After about 40 years of stoppage of sculpting, on the 5th of April, 1722, the Easter that year, Dutch seafarer Jacob Roggerveen unknowingly touched upon the land whereas his voyage was in fact, aimed to locate *Terra Australis*. The islanders are also considered to have a lineage with mythological beliefs. There are many folklores of Rapa Nui religious beliefs. The most prevalent is that *Hotu Matu,* the first settler, the legendary first king of Rapa Nui and Birdman the *Tangata manu* reached the Anakena beach on the island accompanied by his wife and brought with him the so-called navel-of-the-world which was a rounded boulder surrounded by many smaller ones. It was believed by Rapa Nui that all the life ascended from it. Truly the compasses go berserk near this rounded boulder rock mysteriously and it is not explained as yet. Another mysterious rock is *Pu O Hiro* or the trumpet of Hiro, used for sounding the trumpet to summon a gathering or was even believed to attract the fishes to swim to the shoreline. Birdman religion and cult was for the ones who worshipped the creator God *Makemake.* The chiefs of all 12 tribes jointly decided to amicably select a leader to be the 'president' of the entire assemblage of islanders. This selection was made by taking one strong person from each tribe who would swim out to *Moto Nui,* a small island offshore. This coincides with the breeding time of a bird *Sooty Tern, biologicaly Onichprion fuscatus,* an aquatic sea bird belonging to family *Laridae.* It is a bird of the tropical oceans which sleeps on the wing. It returns to the land only to make the nests in or around September and chooses either the coral reefs or hard rock-covered terrains. This selection process involved a competition in which eggs of Sooty Tern would have to be collected by the contestants. The elites of the island were represented by the competent swimmers who would swim to Motu Nui while taking provisions, called *Pora,* under one arm and would try to procure the first egg. The main contestants who sponsored their men actually participating, would wait in village *Orongo.* The first egg finder *Hopu,* would swim back and go to the highest point on Motu Nui and inform his benefactor of the procurement of egg and colour his head red or white after shaving it. He would swim with the egg secured inside a reed basket tied on his forehead back to the land. He would then climb the dangerously steep rocky cliff face without falling down. He would then present the egg to his patron. With this,

Hopu's participation ends and successful contestant, the patron of egg finder *Hopu* would then be declared *'tangata manu'* for the following one year. The entire activity hinged on a variety of threats and was very dangerous. Many would get killed by sharks during the course of swimming or by drowning or would fall from the steep cliff face. The substitutes were easily available and were probably implied ruthlessly. The Birdmen religion was thankfully subdued by the introduction of Christianity on the island. The *Birdman mythology* had in it the most revered God of Rapa Nui, *Makemake.* More recent mythological folklore relates to the story of the epic battle between the *Hanau Epe*, the ones with "long ears", who were a semi-legendary people and *Hanau Momoko*, the ones with "short ears". This epic battle is stated to have taken place between the 16$^{th}$ and 18$^{th}$ centuries, probably in the late 17$^{th}$ century.

The discreet style of monolith statutes resembles what is done in entire Polynesia. The carving is done in fairly flat planes, their faces bearing proud but conspicuously enigmatic expressions. First of all, the human figures would be carved out on the rock surface, and then chipped away until only the image is left. The ratio between head and trunk of the Mo'ai *(Moai)* would be three-to-five and would religiously obey the Polynesian belief in the sanctity of principally, the head. They are invariably given the facial features with heavy brows, elongated noses, a distinct fish-hook-shaped curl in the nostrils and protruding lips in a thin pout. The ears are elongated and oblong in form; noses are prominent. The neck is made truncated with glaring jawlines. The torsos are heavy. The arms are carved in bas relief, resting against the body in various positions. Out of the entire 887, only one Mo'ai *(Moai)* is in a kneeling posture whereas the rest of them do not have clearly visible legs. Some of them are so placed as to be looking like performing the duties of sentinels. They generally face the community but given the narrow width of the island at places, they appear to face the coast. The hemispherical eye sockets were designed to hold the coral eyes with either black obsidian or red

*Scoria* pupils. Abundant red pigment is applied at the human burial sites of a group of buried individuals around the Mo'ai (Moai). It can be deduced that the statues were painted red on the occasion of ceremonies. These burials often surround the statues, suggesting that the Rapa Nui buried their dead with the family's Mo'ai (Moai) which was probably regarded as family's custodian also.

*Ahu Tongariki,* Sculpted by Island Inhabitants: Mo'ai (Moai) in Easter Island; One sporting Pukao, the red headdress

Sculpting by Rapa Nui people was just not a pastime. The statues clearly connoted the symbol of authority and power—religious and social. One can always make out that they were not just symbols for their creators. By the period of the second wave of settlement on the island, the inhabitants had split into several clans who were often in competition with each other, if not in open conflict. In spite of differences and competition, the islanders were unified and shared a religious trust revolving around the cult of ancestors. The rendering of revered ancestors was celebrated in the carving of statues which improved and became much finer over time. They were, in fact, true repositories of sacred spirit of their revered ancestors. The

Polynesian religion included the properly fashioned and ritually prepared, carved stones as well as wooden objects. These objects were believed to be charged by some magical spiritual quintessence known as *'Mana'*. Rapa Nui in fact, also believed in the concept of *'Mana'* which to them was a mystical combination of power, prestige and prosperity. It may thus be perceived that this population was not 'primitive' as described by some European explorers. The Mo'ai *(Moai)* represented its most prestigious ancestors who were believed to bless and bestow 'Mana' on the living leaders, their descendants. Various clans competed with each other by building bigger statues and altars. Dr. George Lee in his book 'The Rock Art of Easter Island' elucidated that creating Mo'ai *(Moai)* was considered to offer a solution to many problems like local skirmishes, crop failure and epidemics which were addressed by devoting more time and efforts to carve bigger statues. The construction of Mo'ai *(Moai)* and *Ahu* served some other important purposes too. Prof. Robert DiNapoli of Department of Anthropology, in the University of Oregon, Willamette Valley, USA identified that *Ahu* were constructed on the coastal spots where freshwater from volcanoes seeped into the ocean. These findings concluded that the erection of *Ahu* by the Rapa Nui society was also used to indicate the presence of freshwater and indicate the particular clan's control over it. According to Prof. DiNapoli, the erection of *Ahu* had a significant ceremonial value also. Mo'ai *(Moai)* occupying the top locus in the society were sculpted to wear a red headdress, called *Pukao* and were virtually the chieftains of their clan. This rock is *Scoria*, a type of *basaltic* lava flow, red in colour ejected as fragments from volcanoes and is very light. This was quarried from another site, *Puna Pau.* Red colour is considered a sacred colour among the Polynesians. Some of the Mo'ai *(Moai)* were painted and decorated in white and maroon colours. Majority of the Mo'ai *(Moai)* stood overseeing their own clan with their back to the ocean, the only exception being the *Ahu Akivi,* a sacred place which has seven Mo'ai *(Moai)* of similar size, facing towards the sea to help travellers to locate the island or waiting for some mythical seafarer to arrive at the island. This peculiarity is also explained with the support of another legend which says that there were seven men waiting for their king to arrive. Generally, the size of the Moʻai *(Moai)* placed on the *Ahu*, proportionately exhibited the religious clout, also a part of what the islanders imbibed as *'mana'*, carried by the chief who commissioned it. Geochemical and mineralogical studies of soil and rocks have been carried out in the island. The spatial distribution of *basalt* quarries together with the use of fine-grained artefacts reveal that Rapa Nui extensively used the stones for various purposes including communal and religious. A most recent study in the year 2019 proposed that ancient people believed that quarrying for Moʻai *(Moai)* might be related to improvement of the soil fertility and thereby critical for food availability. Most Moʻai *(Moai)* had designs carved on the back which had cultural links with

*Ahu Akivi,* A Sacred Place on Easter Island for Inhabitants: Moʻai (Moai) in Easter Island

the island's tattooing art. The DNA analysis of the islanders provided the most reliable conclusion that the Moʻai *(Moai)* were carved by the original inhabitant Rapa Nui and not by a separate group from South America.

Oral traditions, as reported at various places and probably documented also suggest that the Moʻai *(Moai)* were carved either by a distinguished class of professional carvers who were equal to high ranking members of other Polynesian craft guilds or alternatively members of different clans, groups or tribes in the islanders.

The island had completely become a treeless piece of land by about 1200 CE. How such mammoth statues were moved to their respective places is still an intriguing facet and attracted many researchers to conduct experiments. Pollen analysis has clearly established that the island was almost totally forested until about 1200 CE. As per the belief of Rapa Nui the statues 'walked' to their assigned spots. The results obtained from the most recent experiment available in the archaeological records revealed that the statues were harnessed with ropes from two opposite sides, being pulled from one side and loosened from the other, tilting from side to side while pulling forward. This movement appeared just as if the statues 'walked' by *swivelling and rocking* from side to side. This needed synchronised actions of the set of people on three sides in the entire group. They would also recite a rhythmic chant while 'walking' the Moʻai *(Moai)*. The oral history stated in the earlier accounts that a king named *Tuu Ku* moved the statues with the help of the God *Makemake.* Another folklore that came up later mentioned about a woman who lived on the mountains and ordered Moʻai *(Moai)*, about her will to move to various destinations and they obeyed. There is another hypothesis that they used wooden sledges and/or rollers, as well as levelled tracks across the island areas. Yet another view considers the weathered *basalt* boulders, an igneous rock, formed of lava flows that weathers in a typical manner called *onion shell weathering* which results in the formation of almost rounded boulders, laid on more or less evenly dressed tracks leading to various destinations. Apparently making the statutes to 'walk' and 'bubble' along to the destinations, appears most plausible, though ambling distances of 14 or 15 kilometres might have taken many days or maybe months. A Czech Engineer and experimental archaeologist Pavel Pavel together with the renowned Norwegian adventurer Thor Heyerdahl and Kon-Tiki museum experimented with a five-tonnes-and-nine-feet Mo'ai *(Moai)* to validate the possibility of moving them by shuffling motion. It had to be abandoned due to the damage to the statute bases by chipping. However, Thor Heyerdahl estimated that a 20-tonne statute in Easter Island terrain would travel about 100 metres in a day's action. The other school disregarded the idea of such a method of transportation of Mo'ai *(Moai)*. Almost at the same time, a U.S. archaeologist Charles Love also experimented to find the method of transport of the Mo'ai (Moai) but did not publish his findings. Nevertheless, deliberations continue even now.

Initially, the people in Easter Island had a principal leader which was replaced by a warrior class known as *matato'a*. Its symbol was a therianthropic figure comprising half-bird and half-man. It was the Birdman cult, connected to the sacred site in village *Orongo*, situated in the south-western end of the island. Inter-tribal battles began over the worship of ancestry prompted by this cult. During the peak time of the Birdman cult the sculpting of *Mo'ai (Moai)*. had stopped. Instead, 480 petroglyphs were carved by Birdman and God *Makemake* at village *Orongo.* They remained intact because they were carved in solid *basalt* which stood the vagaries of harsh weather over ages. *Orongo* was the venue of Rapa Nui festivities. It is a dangerous landscape, consisting of a narrow ridge with almost a vertical drop of about 1000 feet steep fall into the sea on one side and a deep crater on the other. *Mo'ai (Moai) Kavakava* were smaller with a slender appearance and were made in wood. They spew out a sad appearance. These are thought to have been made after the downfall of the pioneer inhabitants which took place gradually. By 1868, the *Mo'ai (Moai)* placed on *Ahu* were toppled probably a sign of strife, civil war and famine. The French Naval officer *Abel Aubert Dupetit Thouars* was the last to see them erect in 1838.

The petroglyphs encountered in the Easter Island exhibit various fishes including what look like *dolphin* and squid, particularly the one sketched on vertical panels hanging over the cliff edge at *Val Atare*. It matches with the bottlenose *dolphin*, found extensively throughout the waters in Polynesian triangle. The marked difference that draws the attention is that petroglyphs depicting the *Dolphins* do not have 'fish hook' motif drawn close to it while the ones showing sharks and other fishes do. The white coral beach of *Anakena,* has two *Ahu*s: *Ahu Atare* with one *Mo'ai (Moai)* and *Ahu Nao Nao* with seven *Mo'ai (Moai)*. This beach is important as it is one of the two sandy beaches in an otherwise completely rocky coastline and terrain in the island and according to the oral history, received the islander's first king *Hotu Matu'a* who landed here with his wife, the extended family comprising persons from seven tribes and two canoes. Later these people were to constitute seven tribes on the island. Skjdsvold's excavation at Anakena beach unearthed the bones of as many as 13 *Dolphins* together with one harpoon suitable for their hunting. Sources claim that the assemblage of *dolphin* bones was accompanied by human bones also, a narrative that supports the belief that devastated by an acute shortage of food, the Easter Islanders had to resort to the gruesome practice of *cannibalism*.

Apparently, when people planed a feast, they would sail in about ten-to-twelve *canoes* to the open sea and launch the persistent search for *Dolphins*. On spotting a *dolphin* pod, the men will make noise beneath the water by banging rocks or similar objects to scare and confuse the *Dolphins* that will instinctively move away from the source of noise but into the nets of fishermen who had sailed on ahead, in the meantime. *Dolphins* are very sensitive to hearing. Their tiny ears start bleeding and the docile *cetaceans* fall unconscious. When the group gets herded in one area, the *Dolphins* were slaughtered mercilessly for feasting. *Dolphins* and other fishes constituted a significant part of the food for Rapa Nui. Their hunting also needed large *canoes* that in turn could have been made from big timber planks. This construal based on rationality became the basis for ascribing the reason for the debacle of Rap Nui society. Thor Heyerdahl surmised that during about 1100 to 1680 the island had lush vegetation and there were 10 to 15 thousand people. When Jacob Roggerveen arrived, he found there were hardly 3000 persons on the island and virtually no plants around. There are many horrifying incidences related to his visit to the island. One of them is that his group had to shoot 12 islanders to keep them off, probably thinking that they could be cannibals. The island had lost all its plantation presumably to overexploitation of resources delimited over a meagre 163.6 square kilometres of land by Rapa Nui population. There was something that had gone wrong along the way. There was a period of following few decades that the islanders lived in peace. In 1862 a Peruvian incursion to Easter Island was done, resulting in one thousand inhabitants including the king and his family captured for slavery and taken to Peru. This was the case of flagrant '*blackbirding*' during those times which was condemned by many countries. The rest of the islanders were left probably without any festivities and rituals. In Peru also a majority of these slaves died during the next one year or so. At the behest of Tahiti Govt., the Bishop strongly denounced this practice and the embarrassed Peru Govt. had to return the remaining enslaved islanders. The ship returning to Easter Island with leftover islander-slaves faced rampant spread of smallpox leaving just 15 to 20 survivors. Smallpox nearly wiped out the entire Rapa Nui population. This raid for slaves, one of many more from time to time, left the entire Rapa Nui society in complete astonishment. The Peruvians had taken all the able-bodied islanders, among them all the intelligent men and women, particularly the ones who knew how to write and read *rongo-rongo* tablets. Many experts in the field of linguistics have attempted to decode the script but without success. Probably there is no generally accepted method which can be used to read the symbols comprising the dialect. Little could, therefore, be known what was it like in Easter Island before the Europeans came in contact with the people on the island. The *blackbirding* had actually impacted this extremely unpretentious, intelligent and organised society in many ways. Collapse followed resulting from the adverse effect caused due to the impairment of sustainable balance by indiscriminate damage to the trees and vegetation on the island.

Onwards 1868, the entire Easter Island witnessed the collapse of social order on the island. Many of the remaining population accepted the offer to relocate to Tahiti. When the Republic of Chile annexed Easter Island by means of the "*Treaty of Annexation of the island*" on the 9$^{th}$ September 1888, it had a frustrated, disheartened and underprivileged population of only 110 inhabitants. In 1965, however, the islanders were given the citizenship of Chile. The island falls in the territory of *Valparasio region* and constitutes a single commune of the province *Isla de Pascua.*

Realising what Rapa Nui society has been doing by putting in mental and physical exertion and with some muscle, ingenuity, coordination and concerted efforts, one finds it rather challenging to accept the fact that the entire community had to abandon their home, everything that it accomplished over centuries and relocate to a faraway place. Leaving a legacy behind under regress with no future visible must have been very distressing and painful. The Rapa Nui Civilisation attained a level of demographic complexity giving rise to one of the most advanced cultures and technological feats of *Neolithic* societies anywhere in the world. This amazingly paramount society developed, flourished and thrived for about 1000 years or more before collapsing and becoming all but extinct. Why did such an occurrence take place? What actually drove them to the brink? Easter Island leaves us distressed, offers captivating, yet sad story and probably instils a tough lesson as well.

We have voyaged through the passage of the Rapa Nui's birth, the progress, their religious bent, death-defying traditions and creation of astonishing *Mo'ai (Moai)*. But what made them get so disheartened, get reduced so drastically that some of their dead were not even buried and had to ultimately abandon their history at one point of time? There are several models based on various supportings—the physical pieces of evidence and reasonable hypothesis or both. Truly some of them are the result of plain imagination, that have been put forth for the debacle of well-groomed, industrious, religiously bent and socially congenial society, numbering about 15 to 20 thousand in its crowning period inhabiting the island for about 1000 to1100 years at the minimum. The plausible causes assigned in the preceding portion, accountable for the devastation of Rapa Nui do not satisfy the curious and scientific frame of mind.

The alleged self-destruction of islanders, decline and fall of their society has become a subject matter of intense research and debate. What happened to this remarkable society, thriving with such great devotion to their own beliefs? Two diagonally divergent schools of theories have been making rounds for quite some times now. The one that has attained more popularity relates to what is appropriately named the *'ecocide'* theory. American Geographer Jared Mason Diamond has answered the puzzling queries in almost totality in his book *Collapse: How Societies Choose to Fall or Survive, 2005.* In this book of great value, Diamond has devoted one chapter which exclusively focuses on Easter Island. He coined a catchy name for his idea, the *'ecocide' theory*. Diamond's hypothesis is straight, conclusive and can be summarised in limited words. According to Diamond, the islanders had a lasting obsession for carving the *Mo'ai (Moai)* which they did for over three hundred years or more during the middle period of Rapa Nui history. Analysis of pollen fossils had clearly established that the island had a lush jungle of palm trees, *Paschalococos disperta,* also popularly called the Rapa Nui Palm, now extinct, was a native cocoid palm species, exclusively growing in Easter Island. Its pollen record disappeared in 1650 CE and the occurrence is not noticed after that. Its disappearance is attributed to various causes, the growth of the human population which needed building more *canoes* for fishing, more wood for fire and dwelling. Another enemy to the lush palm groves might have been Polynesian rats, arriving on the island as stowaways in the first *canoes* of Polynesian colonists. Once on the island, they found high-quality food, available in plenty in the form of palm nuts. The rats reproduce exponentially whose large population made them more potential for creating havoc to palm saplings and nuts. Polynesian rats, fed on palm nuts and saplings, halted new line of plant generation thus accelerating the demise of palm plantation on the island. Rapa Nui population was also cutting the palm trees to build *canoes* and for use as firewood. The mineral-rich soil of volcanic origin was otherwise a boon for growing plants such as sweet potatoes or yams, other tuberous plants and corn, etc. The bare land on the island with almost complete absence of plants hastened the process of soil erosion which reduced the fertility, consequently affected the cultivation. This must also have increased the surface run-off of rainwater leading to poor or no replenishment of groundwater-table, the sole source of potable freshwater on the island, surrounded by saltwater of Pacific Ocean. Conjointly, all this must have added to the dismay and frustration of inhabitants.

The seafood was also a part of inhabitant's diet. A coherent presumption could be that the access to marine resources varied on the basis of the social and political status of the consumers. People of higher strata might have had greater access to marine food. On observing the topographical setting and varied climatic conditions on Easter Island, certain features appear to belittle the very probability of islanders' access to the fish food from the sea. Argument obtains the support from the topography of the island. The northern end of the island is characterised by steep cliffs, which makes the fishing difficult whereas the southerly part is cooler and might not have been ideal for fishing either. These are not valid arguments, as understandably the *Dolphins* can be hunted only in the open sea with harpoons and needed big *canoes* to bring the hunted ones to the island. *Dolphins* cannot be fished from the coast. When the population on the island reached a few thousand, which initially was just in hundreds, use of timber from palm trees increased and more *canoes* had to be built leading to cutting of larger trees creating a situation when tree cutting exceeded the sustainable yield of the plantation. This resulted in a shortage of *canoes* and those of 'hunted' *Dolphins*. The soil erosion of the treeless island must have enhanced uncontrollably, adversely affecting the cultivation of sweet potatoes or yams, corn (kalo) and other tuberous plants. Food became more and more scarce. This led to internecine warfare among various clans and ultimately to *cannibalism*. The *Neolithic* community of Polynesians inhabiting the barren Easter Island probably could not assess the adverse effect of their particular lifestyle and beliefs, and an incredibly assiduous civilisation collapsed and the remaining few islanders probably got relocated elsewhere, after failingly trying out their hands to adopt *cannibalism*. If this were to be taken as the verdict, the question as to what caused to bring about their own destruction is not at all difficult to answer. As already stated elsewhere, the striking difference in the population count inhabiting the island at the time of Thor Heyerdahl's visit in 1955-56, based on his thoroughly anticipatory estimation, ranged from 10000 to 15000 strong with lush palm groves. It had dwindled to just about 3000 people, when Jacob Roggerveen reached the island on April 5, 1772 and found the land virtually bald with no plantation at all. This dwindling populace and natural resources resulting from the inhabitants' own actions clearly indicate what Diamond had called the 'ecocide'. It was not by any state establishment that the plants were cut so indiscriminately. Instead, the community itself was supervising the matters; it could be very appropriately classified as the self-destruction also. The process of demise materialised so prominently because it had happened in a rather small island 23 kilometres long and 11 kilometres wide stretch of land jutting above the water level of mighty Pacific Ocean. One can easily perceive such unconcealed change in the situation more prominently, compared to the similar one happening on a global scale, over a longer period. One critic of Diamond's *'ecocide'* theory, Dr. Catrine Jarman of Bristol University, England, argued that the process of deforestation might have started with the arrival of first Polynesians settlers who brought along the unwanted Polynesian rats, a voracious rodent who was responsible for the total elimination of palm groves. The islanders continued for centuries even afterwards. Nevertheless, the islanders' contribution to the destruction of plantation on the island cannot be ignored either.

Diamond's *'ecocide'* theory is a glaring instance cited as a cautionary tale against the depletion of natural resources and life virtues, whose exploitation must be undertaken in a manner that would be undoubtedly sustainable with nature and environment. Probably the adverse effect of defying the sustainable living cannot be showcased in a more glorifying manner. The theory of *'ecocide'* as advocated by Jared Diamond through which he has endeavoured to answer several puzzling questions about the Rap Nui debacle is firmly holding its ground, in spite of several criticisms. Diamond's elucidation for its fate of collapse became an exemplary model in the environmental circles and soon became the darling of the lobby, yet certain things were left unattended. In fact, another theory runs parallel to *'ecocide'* theory, by far truly dark and more gruesome that overwhelms the self-destruction hypothesis of a spruce indigenous cultural society. This opinion signposts the expansionism and feudal policies of the colonial regime. It was *blackbirding*.

The popular theory of *'ecocide'* overlooked some historical facts. There has not been any piece of evidence whatsoever, of such a decline in the population. Ethenographic reports provided by oral histories, described warfare on the island which were taken by Thor Heyerdahl as pieces of evidence. He propagated that a civil war which culminated in a battle in 1680, completely eliminated one tribe from the island. The large numbers of obsidian flakes called *"mata'a"* which lay all over the island were interpreted as weapon fragments testifying to this violence. Disagreeing with this, Archaeologist Carl Lipo of Binghamton University, the New York State University had convincingly resolved that these were more likely domestic tools or implements for performing ritual tasks and not the weapons or a part thereof. A few of the human remains showed the signs of injury but none of them so fatal as to have caused any causality. The impression of *cannibalism*, as propounded by a school of scientists, could not garner majority support. A different picture of the historic population on Rapa Nui island got painted. They lived on the island sustainably until European contact that happened due to latter's forays to the island, some of them sociable. As regards the palm Groove plantation, it fell victim to fast-multiplying Polynesian rats, which fed on palm nuts and saplings and halted the growth of new plants. The pollen analysis showed that palm groves disappeared in a short time but humans alone cannot be blamed for it, this opinion proposes. Poor or no yield from farming due to excessive and uncontrollable soil erosion, bad conflicting management of extremely limited freshwater resources did add fuel to the fire. The population of the island went through rough prolonged strife, civil wars and the epidemics and disappeared almost entirely before the arrival of Dutch team with Jacob Roggerveen. Even if some may like to differ in calling this the self-destruction albeit the 'ecocide', it was undeniably a classic case of environmental degradation and the complete destruction of sustainable existence. The slave-raids on the island have a reliable history. In the case of Easter Island, the incidence and suppressive actions against the islanders in 19th century inflicted another derogatory stain on the forehead of the colonial regime. Simple and submissive Rap Nui were targeted for slave-raids many times. The advancement of Peru to invade the island in the 18th and 19th centuries is well-known. In one instance, in December of 1862, eight ships from Peru harboured close to the island and captured some 1000 islanders. This group included the king, his son and the ritual priests. With no priests around, no teachings of ritual practices could be imparted. The captured islanders were sold as slaves in Peru. 90% of them died within one or two years. Benny Peiser of Liverpool John Moores University, Faculty of Science, Liverpool, England in his long, hard-hitting but diagnostic paper in Andrew W. Mellon's Journal Storage *(JSTOR)*, a digital library, is among the strongest proponent of the theory. In a frontal attack he questions the propagator of ecocide hypothesis, "Why has he turned the victims of cultural and physical extermination into the perpetrators of their own demise?" This elaborate analytical work has brought out convincing arguments to decry and proposed that it was Europeans' extravagance for slaves which they thought could be fulfilled by Rapa Nui inhabitants. According to him, the island's environment is a potential paradise and not a wasteland. Thor Heyerdahl in one of his popular books, brought out in 1958 famously asked, "Could these primitive cannibals have been the masters who wrought the classical giant sculptures of aristocratic rile type, which dominated the countryside on this same island?"

The two sides taken up for establishing the credible justification for the complete debacle of Rapa Nui Civilisation together with its allegedly pristine environment in the far past, though largely conflicting, yet in fact partly overlap each other's arena of thinking. Diamond's ecocide hypothesis does not have the support of physical evidence except the pollen grain analysis which proves the presence of extensive palm groves all over the island. Even the timing in terms of period also supports this theory. On the other hand, the *blackbirding*, rampant in those days is a historically recorded fact. Both the schools agree on the point of expertise and commitment the islanders had, to make *Ahu, Mo'ai (Moai)*.and Pukau. Benny Peiser's criticism of Diamond's ecocide, though not totally unfounded, is slightly frenzied. If Rapa Nui did not enforce the ecocide, the population definitely confronted a noticeable disruption in the fine balance of sustainable existence between the living beings and nature on the island. If the islanders and rats didn't do it, who else did? The degradation in various factors of environment—some of them more than the others—characterised the situation on the island. None of these two hypotheses could be completely discarded. Probably the sequence of events has to be taken into account for arriving at what might have happened on the island. When slave-raids began probably much of the ecocide effect was already in place albeit both have to be accepted in their true and specific attire.

European traders, J.B. Dutroux-Bornier and J. Brander's misadventure was in fact, to get the island evacuated of its indigenous population. The slave-raids were conducted repeatedly in the 1870s. The brutal conflicts ensued with shootings resulting in casualties, ending in ecocide. Their houses were burnt and destroyed. The leftover populace of Rapa Nui was deported to Tahiti. The European duo destroyed the existing crop of sweet potatoes. According to Heyerdahl and Ferdon, in 1961, who famously pointed out, "After burning the natives' huts, Dutroux-Bornier had all their sweet potatoes pulled out of the ground three times, to facilitate the persuasion of the starving natives who have thus little hope of surviving on their own island". The systematic destruction of the Rapa Nui social fabrics did play an important role in their demise but the damage caused to the environment and sustainability were important, undoubtedly more than anything else. The absence of any background values for environmental factors is what that could have solved this puzzle with greater accuracy. In the olden times, people perhaps did not bother about the disasters and degradation unaware of the impending danger. Sustainable inhabitation was needed which could have completely averted such an ugly situation.

Anyway, the case of Rapa Nui has to be viewed in terms of what is happening at present and what the future has in store. Another aspect is that these civilisations—the Indus Valley inhabitants, the

Sumerians, the Mayans and the Rapa Nui Polynesians did not know about the others at all; they grew and collapsed in complete isolation, though some of them were contemporaries in terms of the period they prospered. What will happen today? In the integrated global economy scenario, some collapse in one country or region would touch the entire humanity.

These and many more such instances point to one glaring fact, the health and welfare of the environment and its several components are perhaps as necessary as food security to mankind. When a step against nature is ever adopted advertently or otherwise, it leads to some kind of catastrophe, often of the magnitude leading to the complete obliteration of great thriving civilisations that grew through many centuries. This leads us to consider that it is the environment with its innumerable segments that occupies the driver's seat. Imagine the aftermath of a situation when there is very little drinking water left, the freshwater lakes and ponds have shrunk, little oxygen is left for breathing, carbon dioxide and many obnoxious gases predominate the atmosphere, the *ozone layer* is depleted beyond repair, the sea-level has risen to the extent that the human and economic settlements in coastal regions are submerged forcing the population to flee to higher planes thus causing a serious demographic instability, reducing the size of the habitable and farmable land area, leading to a marked shortage of food availability. This is not only depressing but also scary. We do find similarity in situations that faced the people of the Indus Valley, the Sumerians, the Mayans, and the inhabitants of Easter Island. The technologists, however, argue that technology will find respective substitutes. Substitutes for what? For breathable oxygen or habitable and farmable land areas? India and China's foray into automobile adventures are disastrous and the signs are conspicuously visible in cities like Delhi, Bangalore, Beijing and Wuhan. If the automobile majors were to achieve their production targets, probably these two countries alone would require more petroleum products than the world produces presently, producing unfathomable quantity of noxious gases, enlarging the hole already developed in the protective *ozone layer*.

Planet earth is home to millions of species, including humans. Both the natural resources of planet and products of biosphere contribute to the resources that are used to support the human population globally. The global inhabitants are grouped into about 200 independent sovereign states, which interact through commerce, trade, diplomacy, travel and military action. It is, however, congenial to assume that instances narrated above did not occur in a day, month or year. They might have taken a few centuries then but would certainly require much less time now, seeing the human and cattle way of life and their large number on our planet. In an attempt to retrieve the lost treasure of our past, safeguarding the rich legacy from those who out of greed and the sense of 'pseudo-self', intend to minimise the value of the fact that all of us, across the planet have a shared past. We venture out to proclaim that although the norms of our culture may have stark differences now, the hitches, the intricacies, and the respective way outs are virtually the same as those of people in the times of Indus-Saraswati, the Sumerians, the Mayans and even the Rapa Nui.

The universe formed, expanded, grew; the sun, the earth, the moon, the milky ways, the galaxies, the stars and the nebulae formed. The earth became habitable and gave rise to an enormous variety of interesting creatures and ultimately humans took over. This has been a mind-boggling phenomenon and has taken time—billion and billions of years. As the environment became more unpredictable, bigger brains helped our ancestors survive. They had largely become vegetarian though not exclusively. They spread across the length and breadth, inhabiting every nook and corner of the planet. They made scientific inventions to help lead a more and more comfortable life. They made sophisticated weapons capable of destroying everything at the flick of a finger.

All this, however, has brought off a transformation of the order which is potent enough to bring off unintended consequences for other species as well as for ourselves, creating new challenges for survival. Not a piece of mouth flapping but a daring disclosure, "the grandeur of the planet and nature is under a terminal threat". The human is the most intelligent of the lot in the living world and thus the onus of its upkeep inevitably rests on its capable and deserving shoulders. Undeniably, these bizarre illustrations are worthy of conveying emphatically that the safety and sound health of the environment is almost as essential as the existence of the sun and the solar system.

# Bibliography of References & Further Readings

| AUTHOR/ RESEARCHER | YEAR | DETAILS OF WORK |
|---|---|---|
| *Abrams, Elliot M.* | 1994 | How Maya built their world: Energetics and Ancient Architecture. *Austin, Texas, University of Texas Press.* ISBN 978-0-292-70461-9 |
| *Adams, Richard E. W.* | 2005 (1977) | Prehistoric *Mesoamerica (3$^{rd}$ ed.). Norman, Oklahoma: University of Oklahoma Press.* ISBN 978-0-8061-3702-5 |
| American Academy of Arts Sciences | 1780-2010 | Book of Members |
| Article from the Jewish Telegraph Agency archive | December 13, 1993 | Astronaut Spins More Than Telescope |
| *Baker, P. E.; Buckley, F.; Holland, J. G.* | 1974 | *Petrology and geochemistry of Easter Island". Contributions to Mineralogy and Petrology. 44 (2): 85–100 Bibcode:1974CoMP...44...85B. doi:10.1007/BF00385783.* |
| *Barraclough, Geoffrey; Stone, Norman* | 1989 | The Times Atlas of World History. *Hammond Incorporated.* ISBN 9780723003045 |

*Continued…*

| Barthel, Thomas S. | 1974 | The Eighth Land: The Polynesian Settlement of Easter Island (1978 ed.), University of Hawaii |
|---|---|---|
| Basham, A.L. | 1967 | The Wonder that was India: Sidgwick & Jackson, London |
| Bhabha, Homi Jehangir | 1935 | On cosmic radiation and the creation and annihilation of positrons and electrons |
| Biographical Memoirs of Fellow of the Royal Society | 1894 (January) | Satyendra Nath Bose *rsbm 1975.000* |
| Bottéro, Jean, André Finet, Bertrand Lafont, and George Roux. | 2001 | Everyday Life in Ancient Mesopotamia. *Edinburgh: Edinburgh University Press, Baltimore: Johns Hopkins University Press.* |
| Bowler, Peter | 1977 | Edward Drinker Cope and the Changing structures of the evolutionary theory |
| Boyd, Robert; Silk, Joan B. | *2003* | How Humans Evolved (3$^{rd}$ ed.). *New York: W.W. Norton & Company. ISBN 978-0-393-97854-4. LCCN 2002075336.* |
| Bradshaw Foundation | 2015. | Sentinels in Stone – The Collapse of Easter Island's Culture |
| Braswell, Geoffrey E. | 2014 | *The Maya and their Central American Neighbours: Settlement patterns, architecture, hieroglyphic texts, and ceramics. Oxford, UK and New York: Routledge. ISBN 978-0-415-74487-4.* |
| Bridget Allchin, Raymond Allchin | 1982 | The Rise of Civilisation in India and Pakistan, *Cambridge University Press* |
| Brooke, John L. | 2014 | *Climate Change and the Course of Global History: A Rough Journey, Cambridge University Press, ISBN 978-0-521-87164-8* |

| Brown, Lester R. | 2003 | Eco-Economy Building an Economy for the Earth ISBN 81 250 2203 1 |
|---|---|---|
| Bulliet, Richard W | 2016 | The wheel: Inventions and Reinventions |
| Burbidge, E.M., Burbidge, G.R., Fowler, W.A. And Hoyle, F | 1957 | Synthesis of the Elements in Stars |
| Carter, Nicholas P | 2014 | Sources and Scales of Classic Maya History In Kurt Raaflaub (ed.). Thinking, Recording, and Writing History in the Ancient World. New York: Wiley-Blackwell |
| Chatterjee, Anirban; Ray, Jyotiranjan S.; Shukla, Anil D.; Pande, Kanchan | 2019 | "On the existence of a perennial river in the Harappan heartland". *Scientific Reports*. 9 (1): 17221. ISSN 2045-2322. |
| Christie, Jessica Joyce | | *Maya Palaces and Elite Residences: An Interdisciplinary Approach.* Austin, Texas: University of Texas Press. ISBN 978-0-292-71244-7. |
| Clift, Peter D.; Carter, Andrew; Giosan, Liviu; Durcan, Julie | 2012 | U-Pb zircon dating evidence for a Pleistocene Saraswati River and capture of the Yamuna River Geology. 40 (3): 211–214. |
| Coe, Michael D. | 1999 | The Maya (Sixth ed.). *New York: Thames & Hudson.* ISBN 978-0-500-28066-9 |
| Col Dr. Laurence Waddell | 1927 | Aryan Origin of the alphabet and Sumer-Aryan Dictionary |
| Cowlishaw, G. & Dunbar, R. | 2000 | Primate Conservation Biology. *Chicago: University of Chicago Press.* ISBN 978-0-226-11637-2. |

*Continued...*

| Crawford, Harriet E. W. | 2004 | Sumer and the Sumerians. *Cambridge: Cambridge University Press.* |
|---|---|---|
| Dart, Raymond Arthur | 1959 | Adventures with missing links; *Harper, New York* |
| Demarest, Arthur | 2004 | Ancient Maya: The Rise and Fall of a Forest Civilisation. *Cambridge, UK: Cambridge University Press. ISBN 978-0-521-53390-4.* |
| *Demarest, Arthur Andrew; Prudence M. Rice & Don Stephen Rice* | 2004 | The Terminal Classic in the Maya Lowlands: Collapse, Transition, and Transformation. *Boulder, Colorado: University Press of Colorado. ISBN 978-0-87081-739-7* |
| *Diamond, Jared* | *1999* | Guns, Germs, and Steel: The Fates of Human Societies. *New York: W.W. Norton & Company. ISBN 978-0-393-31755-8. LCCN 2005284124* |
| *Dr. Goodall, Jane* | 1971 | *In the Shadow of Man Boston: Houghton Mifflin* |
| *Dr. Goodall, Jane* | 1986 | The Chimpanzees of Gombe: Patterns of Behaviour *Cambridge, MA: Belknap Press of Harvard University Press.* |
| *Dr. Goodall, Jane* | 1970 | My Friends the Wild Chimpanzees: *Washington, DC: The National Geographic Society.* |
| *Dr. Goodall, Jane, McAvoy: Hudson, GAll* | 2005 | Harvest for Hope: A guide to Mindful eating *New York, Warner Book* |
| Dubois, Marie Eugene Francois | 2012c & 2018 | The Climates of the Geological Past and their Relation to the Evolution of the sun |
| Encyclopaedia Britannica | | Satendra Nath Bose- *Bengali physicist* |
| Englert, Sebastian F. | 1970 | Island at the Center of the World. *New York: Charles Scribner's Sons* |

| | | |
|---|---|---|
| Erickson, Jon D.; Gowdy, John M. | 2000 | Resource Use, Institutions, and Sustainability: A Tale of Two Pacific Island Cultures. *Land Economics.* 76 (3): 345–354. doi:10.2307/3147033. JSTOR 3147033. |
| Farmelo, Graham | 2010 | The Strangest Man: The Hidden Life of Paul Dirac, Quantum Genius *Faber & Faber ISBN 9780571222780* |
| Ferguson, Kitty | 1991 | Stephen Hawking Quest For A Theory of Everything Franklin Watts *ISBN 0-553-29895-X* |
| *Fitzsimmons, James L* | 2009 | *Death and the Classic Maya Kings. Austin, Texas: University of Texas Press. ISBN 978-0-292-71890-6* |
| Foster, Lynn | 2002 | Handbook to Life in the Ancient Maya World. *New York: Oxford University Press. ISBN 978-0-19-518363-4* |
| Frontiers of Astronomy | 1955 | *Heinemann Education Books Limited Archived from Internet; Archive Harper Collins, ISBN 978-0060027605* |
| Garber, James | 2004 | *The Ancient Maya of the Belize Valley: Half a Century of Archaeological Research. Gainesville, Florida: University Press of Florida. ISBN 978-0-8130-2685-5.* |
| Gingerich, Owen | 2004 | The book Nobody Read: Chasing the Revolutions of Nicolas Copernicus, *New York: Walker. ISBN 0-8027-1415-3* |
| Gittelman, Rachel M.; Schraiber, Joshua G.; Vernot, Benjamin; Mikacenic, Carmen; Wurfel, Mark M.; Akey, Joshua M. | 2016 | Archaic Hominin Admixture Facilitated Adaptation to Out-of-Africa Environments. *Current Biology.* 26 (24): 3375–3382 |

*Continued...*

| Goldhaber, Gerson | 2009 | The Acceleration of the Expansion of the Universe: A Brief Early History of the Supernova Cosmology Project (SCP) *Proceedings of the 8th University of California, Los Angeles, Dark Matter Symposium* |
|---|---|---|
| Goodman, Billy | 2002 | Big days for the Big Bang *Princeton Alumni Weekly* |
| Gordon, Dave | March, 2017 | Practising Judaism in Space- Jewish astronauts reflect upon their time in outer space *Community Magazine, Brooklyn* |
| *Habib, Irfan* | *2015* | *The Indus Civilisation. Tulika Books. ISBN 978-93-82381-53-2.* |
| Harrison, David | *2009* | *Nature* (British Journal); Springer Nature, London(July, 2009) |
| Haun, Beverley | 2008 | *Inventing Easter Island Archived 12 April 2016 at the Wayback Machine. University of Toronto Press. p. 8. ISBN 0-8020-9888-6(* |
| Heyerdahl, Thor; Edwin N. Ferdon Jr. (eds.). | 1961 | The Concept of *rongo-rongo* among the Historic Population of Easter Island. *Stockholm:* |
| Hoyle, F | 1962 | Astronomy, A history of man's investigation of the universe: *Crescent Books, Inc., London* |
| Hoyle, Fred | 1946 | The Synthesis of the Elements from hydrogen *Monthly notices of the Royal Astronomical Society* |
| Hoyle, Fred & Hoyle Geoffrey | 1982 | The Frozen Planet of Azuron |
| Hoyle, Fred & Hoyle Geoffrey | 1982 | The Energy Pirates Ladybird Books |
| Hoyle, Fred & Hoyle Geoffrey | 1982 | The Planet of Death Ladybird Books |

| | | |
|---|---|---|
| Hoyle, Fred & Hoyle Geoffrey | 1982 | The Giants of Universal Park |
| Hoyle, Fred & Wickramansinghe, N.C., | 1984 | From Grains to Bacteria *University College Cardiff Press, INBN 0-906449-64-2,* |
| Hubble, Edwin | 1926 | Extragalactic nebulae, Astrophysical Journal **ISSN**: 0004-637X (print); 1538-4357 (web) |
| *Hunt, T* | 2006 | *Rethinking the Fall of Easter Island American Scientist. 94 (5): 412. doi:10.1511/2006.61.1002. Corrections in radiocarbon dating suggests that the first settlers arrived from other Polynesia islands around 1200 A.D.* |
| *Hunt, Terry; Lipo, Carl* | 2011 | *The Statues that Walked: Unravelling the Mystery of Easter Island. Free Press. ISBN 978-1-4391-5031-3* |
| *Johanson, Donald; Edey, Maitland* | 1981 | *Lucy, the Beginnings of Humankind. St Albans: Granada. ISBN 978-0-586-08437-3.* |
| John Montgomery | 2001 | Tikal: An Illustrated History of the Mayan Capital *published by Hippocrene Books* |
| Jonathan Bird | | Jonathan's Blue World *Video films on Maya and others* |
| *Jones, David T.* | 2007 | Easter Island, what to learn from the puzzles? *American Diplomacy. Archived from* the original *on 28 November 2007.* |
| Jones, Grant D. | 1998 | The Conquest of the Last Maya Kingdom. *Stanford, California: Stanford University Press. ISBN 978-0-8047-3522-3* |
| *Kalyanaraman, S., ed.* | 2008 | *Vedic River Saraswati and Hindu Civilisation, Aryan Books International, PP.308, ISBN 978-81-7305-365-8* |

*Continued...*

| Kappeler, Peter M. | 1998 | Nests, Tree Holes, and the Evolution of Primate Life Histories American Journal of Primatology. 46 (1): 7–33. |
|---|---|---|
| Kenoyer, Jonathan Mark | 1998 | Ancient cities of the Indus Valley Civilisation. Oxford University Press. ISBN 978-0-19-577940-0. |
| Kramer, Samuel Noah | 1988 | In the World of Sumer: An Autobiography. Wayne State University Press. p. 44. ISBN 9780814321218. |
| Kramer, Samuel Noah | 1963 | The Sumerians: Their History, Culture, and Character. The Univ. of Chicago Press. ISBN 978-0-226-45238-8. |
| Kramer, Samuel Noah | 1983 | The Sumerian Deluge Myth: Reviewed and Revised", Anatolian Studies, 33: 115–121, doi:10.2307/3642699, JSTOR 3642699 |
| Kramer, Samuel Noah | 2007 & 1972 | Sumerian Mythology: A Study of Spiritual and Literary Achievement in the Third Millennium B.C. Forgotten Books. pp. 1–182. ISBN 9781605060491. |
| Kramer, Samuel Noah | 2010 & 1963 | The Sumerians: their history, culture, and character. University of Chicago Press. pp. 1–372. ISBN 9780226452326. |
| Lahiri, Nayanjot, | 2000 | The Decline and Fall of the Indus Civilisation. Permanent Black. ISBN 978-81-7530-034-7. |
| Lal, B.B. | 1997 | The Earliest Civilisation of South Asia (Rise, Maturity and Decline). |
| Lal, B.B. | 2002 | The Saraswati flows on. |
| Larsen, A. and Simpson Jr., G.F. comment to Rull et al | 2013 & 2014 | Challenging Easter Island's Collapse: the need for interdisciplinary Synergies. Frontiers in Ecology and Evolution 15 dol: 10.3389/fevo.2014.00056.D.F. |

| Leakey, Richard E. | 1994 | The Origin of Humankind. *Science Masters Series. New York: Basic Books. ISBN 978-0-465-03135-1. LCCN 94003617.* |
|---|---|---|
| Leaky, Richard Erkine Frere | 1973 | Evidence for an Advanced Plio-Pleistocene Hominid from East Rudlf, Kenya, *Nature PMID 4700897* |
| Lee, Georgia | 1992 | The Rock Air of Easter Island. Institute of Archaeology. *UCLA* |
| Lee. Georgia. | 1990. | Two Easter Island Legends that Relate to Petroglyphs. *Archae, Vol. 1, Fall.* |
| Lemonick, Michael D | 2003 | Echo of the Big Bang, *Princeton University Press. Pronceton, New Jersey, USA ISBN 0-691-12242-3* |
| Lewis, Raymond J. | 1994 | Review of Rapanui; Tradition and Survival on Easter Island *Archived 20 May 2013 at the Wayback Machine* |
| *Mackenzie, Donald Alexander* | *1927* | *Footprints of Early Man. Blackie & Son Limited.* |
| Madougall, I; Brown, F.H.; Vasconcelos, P.M.; Cohen, B.E | 2012 | New single crystal 40 Ar/39 Ar ages improve time scale for deposition of the Omo-Group, Omo-Turkana Basin, East Africa. *Journal of the Geological Society, London* |
| Maemoku, Hideki;Shitaoka, Yorinao; Nagatomu, Tsunedo; Yagi, Hiroshi | 2012 | Geomorphological Constraints on the Ghaggar River Regime During the Mature Harappan Period *Online ISBN 9781118704325 & Print ISBN 9780875904887* |
| Mahanti, Dr. Subodh | | Satyendra Nath Bose, The Creator of Quantum Statistics *IN: Vigyan Prasar* |
| *Maisels, Charles Keith* | 1993 | The Near East: Archaeology in the "Cradle of Civilisation". *ISBN 978-0-415-04742-5.* |

*Continued...*

| Marshall, John (editor) | 1931 | *Mohenjo-Daro and the Indus Civilisation: Being an Official Account of Archaeological Excavations at Mohenjo-Daro Carried Out by the Government of India Between the Years 1922 and 1927.* London: Arthur Probsthain. ISBN 978-81-206-1179-5. |
|---|---|---|
| Masson, Marilyn A. | 2012 | Maya collapse cycles *Proceedings of the National Academy of Sciences of the United States of America. 109 (45): 18237–38. JSTOR 41829886.* |
| Masson, Marilyn A.; Carlos Peraza Lope | 2014 | "Militarism, Misery and Collapse." In Marilyn A. Masson; Carlos Peraza Lope (eds.). Kukulcan's Realm: Urban Life at Ancient Mayapán. *Boulder, Colorado, US: University Press of Colorado.* ISBN 978-1-60732-319-8. |
| McIntosh, Jane | 2008 | *The Ancient Indus Valley: New Perspectives. ABC-CLIO.* ISBN 978-1-57607-907-2. |
| Meave Leaky: Bios | Archived from Internet | Department of Anthropology, Turkana Basin Institute, Stony Brook University |
| Mehta, Jagdish; Rechenberg, Helmut | 2001 | The historical Development of Quantum Theory *Springer Science & Business Media* ISBN 9780387951805 |
| Michael Coe | 2011 | "The Maya" *published by Thames and Hudson* |
| Miller, E. R.;. Gunnell, G. F.; Martin, R. D | 2005 | Deep Time and the Search for Anthropoid Origins. *American Journal of Physical Anthropology.* 128: 60–95. doi:10.1002/ajpa.20352 |
| Misra, Dr. Nikhil C | 2017 | Natural Resources, Eco-economy and Environment: *the Belligerent Insight, International Conference* |

| | | |
|---|---|---|
| Misra, Dr. Nikhil C | 2016 | A Narrative on the Past Records of Minerals, Mining, Humans and Environment *Recent Advancements in Mineral and Water Resources, Excellent Publishers, New Delhi ISBN 978-93-84935-84-9* |
| Mitton, Simon | 2011 | Fred Hoyle; A life in Science *ISBN-10: 0521189470* <br> *ISBN-13: 978-0521189477 Cambridge University Press* |
| Mukherjee, Ashok | 2001 | *Rigvedic Saraswati: Myth and Reality, Breakthrough Science Society* |
| Mukunda, N. | 1987 | The Life and work of Paul Adrien Maurice Dirac *Recent Development in Theoretical Physics, World Scientific MR 935624* |
| NASA, archived from the original | 2011 | Biography of Edwin Hubble (1889-1953) |
| Naked Science | Several | Informative and analytical videos on various subjects prepared and shown, streaming on various web sites |
| Nicholas Copernicus (Polish: Mikolaj Kopernik) Translated with commentary by Rosen, Edward | 1543 (Original) 1978 (Translated in English and Published in Warsaw) | *Original*-De revolutionibus orbium coelestium Translated in English *"On the Revolution of the Heavenly Spheres"*: English Translation: Johns Hopkins University Press, 1992, INBN 0-8018-4515-7 |
| Otto Schoetensack | 1908 | The lower jaw of the Homo heidelbergensis out of sands of Mauer near Heidelberg. Leipzig: *Wilhelm Engelmann.* |

*Continued...*

| Owen, James | 2007 | "Modern Humans Came Out of Africa, 'Definitive' Study Says". *National Geographic News*. Washington, DC: National Geographic Society |
|---|---|---|
| Parpola, Asko | 2015 | *The Roots of Hinduism. The Early Aryans and the Indus Civilisation. Oxford University Press.* |
| Penzias, Arno Allan | 2011 | I E E E Global History Network, *IEEE* |
| Penzias, Arno Allan & Wilson, R.W. | 1965 | A Measurement of Excess Antenna, Temperature at 4080 Mc/s" *Astrophysical Journal* |
| *Polynesian Voyaging Society* | 2010 | *The Voyage to Rapa Nui 1999–2000* |
| *Portal RapaNui, the island's official website, archived from the original* |  | Welcome to Rapa Nui – Isla de Pascua – Easter Island |
| Possehl, Gregory L. | 2002 | The Indus Civilisation: A Contemporary Perspective. Rowman Altamira. ISBN 978-0-7591-1642-9. |
| Press Information Bureau, Government of India | 2013 | Mythical Saraswati River |
| Princess of Asturias Foundation | 29 October, 2013 | Peter Higgs, Francois Englert and European Organisation for Nuclear Research CERN |
| Professor (Dr.) George Bay | 2007 | National Science Foundation (NSF) project on study of agricultural practices and Architectural construction of ancient Maya Civilisation in Yucatan, México. *Department of sociology and anthropology and Chisholm Foundation Chair of Arts and Sciences at Millsaps College ISBN 0-226-00544-5* |

| | | |
|---|---|---|
| Rao, Shikaripura Ranganatha | 1991 | Dawn and Devolution of the Indus Civilisation. New Delhi: Aditya Prakashan. ISBN 978-81-85179-74-2. |
| Reade, Julian E | 2008 | The Indus-Mesopotamia relationship reconsidered (Gs Elisabeth During Caspers). Archaeopress. ISBN 978-1-4073-0312-3. |
| American documentary series; PBS (Public Broadcasting System) - an associate of National Geographic Channel video films | ——— | Informative and analytical videos on various subjects prepared and shown, streaming on various web sites. |
| Redd, Nola Taylor | 2018 | Famous astronomers/List of Great Scientists in Astronomy SPACE.com |
| Reese Riley of the USA | 2018 | Most influential Astronomers of All time Futurism Jerric Ventures LLC |
| Restall, Matthew; Florine Asselbergs | 2007 | Invading Guatemala: Spanish, NAhua, and Maya Accounts of the Conquest Wars. University Park, Pennsylvania: Pennsylvania State University Press. ISBN 978-0-271-02758-6. |
| Richard Leventhal | 2012 | Carlos Chan Espinosa and Cistina Coc 2012 edition of Expedition magazine. |
| Robert E. & Carolyn Krebs | 2003 | Groundbreaking Scientific Experiments, Inventions and Discoveries of the Ancient World |
| Robert McCormick Adams | 1981 | Heartland of Cities: Surveys of Ancient Settlement and Land use on the Central Floodplain of the Euphrates, University of Chicago Press, Chicago, USA |

*Continued...*

| | | |
|---|---|---|
| Rowe, N. | 1996 | The Pictorial Guide to the Living Primates. *Pogonias Press. pp. 4, 139, 143, 15 185, 223. ISBN 0-9648825-0-7* |
| Rui Zhang; Yin-Qiu Wang; Bing Su | *2008* | Molecular Evolution of a Primate-Specific micro-RNA Family" (PDF). Molecular Biology and Evolution. *25 (7): 1493–1502. doi:10.1093/mol bev/msn094. ISSN 0737-4038.* |
| Russell, Joanna | 2009 | *Rapamycin* – Introduction. *Archived from the original on 26 July 2009.* & *Rapamycin* Extends Longevity in Mice. *Archived from the original on 29 May 2010.* |
| Sankaran, A.V. | 1999 | The ancient river lost in the desert *Current Science, 77 (8) JSTOR 24103577* |
| Sarkar, Anindya et.al. | 2016 | "Oxygen isotope in archaeological bioapatites from India: Implications to climate change and decline of Bronze Age Harappan Civilisation". *Scientific Reports* volume 6, *Article number: 26555 (2016)* |
| Schomp, Virginia. | 2004 | Ancient Mesopotamia: The Sumerians, Babylonians, and Assyrians. |
| *Shaffer, Jim G.* | *1999* | *Migration, Philology and South Asian Archaeology". In Bronkhorst; Deshpande (eds.). Aryan and Non-Aryan in South Asia. Cambridge: Harvard University, Dept. of Sanskrit and Indian Studies. ISBN 978-1-888789-04-1* |
| Shah, Niranjan | 2012 | Indian Origins of Ancient Civilisations Egyptian, Sumerian, Greek, Chinese |
| Sharer, Robert J.; Loa P. Traxler | 2006 | The Ancient Maya (6th, fully revised ed.). *Stanford, California: Stanford University Press. ISBN 978-0-8047-4817-9* |

| | | |
|---|---|---|
| Shepardson, Britton | 2013 | Moai: A New Look at Old Faces. *Santiago: Rapa Nui Press. ISBN 978-9569337000.* |
| Simpson Jr | 2014 | Review of Rapa Nui's Geodynamic, Volcanic and Geologic Evolutio *Apuntes de la Bililioteca William Mulloy 3: 1-30* |
| Simpson Jr. D.F. and Dussubieux, L | 2018 | A collapse narrative? Geochemistry and spatial distribution of *basalt* quarries and fine-grained artefacts reveal communal use of stone on Rapa Nui (Easter Island). *Journal of Archaeological Science: Reports* 18:370-385 |
| Skjlsvold, A., | 1994 | Archaeological Investigations at Ana-kena, Easter Island. *The Kon-Tiki Museum* |
| Smithsonian Institution | ---------- | Archaeology, Ecology and Culture. *Washington, D.C.: Smithsonian Institution Press. ISBN 978-0-7141-2504-6.* |
| Sperge, David Nathaniel | 1982 | The jolly red giant-type evolved stars and their evolution to planetary nebulae *Princeton, New Jersey, Department of Astrophysical Sciences* |
| Spergel, D.N. & Steinhardt, P.J | 2000 | Observational Evidence for Self-Interacting Cold Dark Matter |
| Stephen Bertman | 2002 | Handbook to life in Ancient Mesopotamia |
| *Sullivan, Herbert P.* | *1964* | *A Re-Examination of the Religion of the Indus Civilisation History of Religions 4 (1). JSTOR 1061875.* |

*Continued...*

| | | |
|---|---|---|
| *Britannica Online Encyclopaedia.* Britannica.com | | Sumer (ancient region, Iraq) |
| Thapar, Romila | 2004 | *Early India: From the Origins to AD 1300.* University of California Press. ISBN 978-0-520-24225-8. |
| *The Guardian.* Associated Press | 2018 | "Scientists discover ancient Mayan city hidden under Guatemalan jungle" |
| The Hindustan Times | 2020 (July) | Hunt for mythical Saraswati River a test of history and science |
| The Indian Express | 17 March 2009 | Indian Scientists discover three new species of bacteria |
| The Nature of the Universe | 1950 | A series of broadcast lectures *Basil Blackwell, Oxford* |
| Wali, Kameshwar | 2009 | Satyendra Nath Bose: his life and times ( selected works with commentary), *Singapore: World Scientific* ISBN 978-981-279-070-5 |
| Walker, A; Leaky, R.E.F. | 1978 | The Hominids of East Turkana, *Scientific American JSTOR 24960354 PMID 98842* |
| Weldon Lamb of New México State University | 2017 | "The Mayan calendar: A Book of Months", *published by University of Oklahoma Press,* 2017 |
| Willford, John Noble | September, 2007 | New Fossils offer Glimpse of Human Ancestors, *New York Times* |
| *Williams, B. A.; Kay, R. F.; Kirk, E. C.* | | *New perspectives on anthropoid origins Proceedings of the National Academy of Sciences of the United States of America. 107 (11)4797-4804. Bibcode:2010PNAS..107.4797W* |

| | | |
|---|---|---|
| Willoughby, Pamela R. | 2005 | *Palaeoanthropology and the Evolutionary Place of Humans in Nature. International Journal of Comparative Psychology. 18 (1):* 60–91. ISSN 0889-3667 |
| Witschey, Walter R. T.; Clifford T. Brown | 2012 | Historical Dictionary of Mesoamerica. *Plymouth, Devon, UK: Scarecrow Press.* ISBN 978-0-8108-7167-0 |
| Wright, Rita P | 2009 | *The Ancient Indus: Urbanism, Economy, and Society. Cambridge University Press.* ISBN 978-0-521-57219-4 |
| | 2017 | Ancient History Encyclopaedia |
| Gover, Jerry | 2019 | The sword of Schule |
| McIntosh, Jane | 2008 | The Ancient Indus Valley: New Perspectives. ISBN 978-1-57607-907-2 |
| https://tamilandvedas.com/ | 2012 | 2012/04/04/karnataka-indus-valley-connection/ |
| Ancient ruins of Mayan Pyramids--Tazumal | | https://ancientmayanruins.com/tazumal/ |
| Franke, Dr. Richard W | 2019 | Montclair State University Department of Anthropology Anth 140: Non-Western Contributions to the Western World |

# Index of Words

**A**
a thin-walled, high-vaulted skull, 83
A Toroidal LHC Apparatus CMS combine (*ATLAS*), 26
ability of vanishing, 164
Aboriginal people, 48
accumulation of salt, 154
acheulean tools, 76
Adab, 146
Adaptive Radiation, 57
Adverse impact on water resources, 114
aesthetic damage, 101
African Raymond Dart, 75
African Siamins, 59
Agade, 146
agricultural lands, 101
Agriculture activities, 114
*Ahu*, 172, 175, 178-180, 182-183, 191
*Ahu Akivi,* A Sacred Place, 179-180
*Ahu Tongariki,* 175, 178
*Ahu* Vinapu, 175
Ainu people, 81

Akkadian, 143, 145, 148, 153
Akshak, 146
al-Ubaid, 144
Alamgirpur, in Meerut district, 123
Albert Einstein, 20
Alfred Russell Wallace, 52
alignment of the sun, 172
Allahdino in Karachi district, 124
Altai mountains of Siberia, 86
altar-like platforms, 172
Amavasya, 44
amazingly paramount society, 185
American archaeologist Walter Fairservis, 137
American Geographer Jared Mason Diamond, 186
American Institute of Vedic Studies, 135
Amorite migration, 146
Amphibians, 39, 57
*Amu Dariya, the Oxus River*, 129
Anakena beach, 176, 183
anatomically modern *Homo sapiens*, 83

Ancient Egyptian Civilisation, 119
ancient Yamuna, 121
Andes range, 172
Andrew W. Mellon's Journal Storage (JSTOR), a digital library, 191
Androbinan Nebula, 20
Angiosperms, 57
anorthoclase crystals, 71
Anoxia, 37, 40
Anoxic, 37-38
antiquity, 3, 9, 31, 33, 35, 37, 39, 41, 43, 45, 47, 49-50, 88, 119, 125
Archaeologist Carl Lipo of Binghamton University, the New York State University, 190
archaic species, 75
Archosaurs, 38-39
Arctic ocean, 49
Ardi, 65
*Ardipithecus*, 63, 65-67, 117
*Ardipithecus* group, 65-67, 117
*Ardipithecus* kadabba, 65-66
*Ardipithecus* ramidus,, 65-66
Argentina, 37, 58
*aringa ora ata tepuna*, 175
Arthropods, 38, 56
Āryabhaṭīya, 28
Aryabhatt, 8
Aryan, 131, 134-136, 199, 203, 208, 210
Asia was the cradle of humanity and not Africa, 87
Assyrilogist, 144
Asteroids, 8, 20, 27, 29, 34, 46-47
astrology, 151

astronomically oriented, 172
atmosphere, 8, 21, 27, 29, 32, 34-35, 38, 42, 47, 49, 56, 93-99, 101-102, 110-113, 154, 165, 193
Aurignacian culture, 81
auspicious symbol of "Swastik", 128
Australia, 35, 47-48, 84, 91, 114
*Australopithecus*, 63, 68-69, 71, 73, 85, 117
*Australopithecus* afarensis, 68
*Australopithecus* africansis, 68
*Australopithecus* anamensis, 68
*Australopithecus* garhi, 68, 71
Austrlopithecus afarensis, 75
Avocado, 9, 155
Axchiltan Chiapas in México, 159
Azolla ferns, 49
*Aztec*, 119, 155, 158, 161

**B**

Babar Kot in Saurashtra district, 123
Baboons, 60
Babylon, 146-147, 150
Babylonia, 9, 143
*bacterium Streptomyces hygroscopicus*, 170
*Bactria Margiana Archaeological Complex* (BMAC), 139
Bad-Tibira, 146
Baghdad to the Persian Gulf, 143
Balakot in Lasbela district of Balochistan province, 124
balance between rainfall and evaporation, 166
Bananas, 173
Bargaon, Hulas in Saharanpur, 123

Baror in Sri Ganganagar district of Rajasthan state in India, 123
*Basalt*, 31, 170, 174, 180-182, 211
*basalt* boulders, 181
*Basalt* to peralkaline Rhyolites, 170
base substitutions, 50
Beas, 136-137
Belize, 9, 155-156, 159-161, 166, 201
Bell laboratory, 22
Benny Peiser of Liverpool John Moores University, Faculty of Science, Liverpool, England, 191
Berkeley Earth, 111
Berkeley, California, 111
Bernard Ngeneo, 73
beside man, 71
Bhagatrav, 123, 129
biblical book of genesis, 143
*Big Bang Theory*, 8, 20-21
Biodiversity, 36, 40, 48, 91, 117
Biosphere, 31-32, 93-94, 97, 194
Biotic factors, 95
Birdman mythology, 177
Birdman the *Tangata manu*, 176
Bishop strongly denounced, 184
bizarre illustrations, 195
Black and White Ruffed Lemur, 59
Black hole, 22
black shale, 38
Black Skull, 72
*Blackbirding*, 184-185, 189, 191
bleeding, 183
blue hole, 166

blue-green algae, 36, 48
Bonampak, 159
Bonobos, 60-61, 63, 66, 116-117
book "The Maya", 161
Borahshi, 146
Borsippa, 146
Bose-Einstein statistics, 26
Boskop Man, 87
Bouri formation, 71
brachiopod temnospondyl, 39
Bredford Dome, 35
brick columns, 150
Brigitte Senut, 66
Brink, 185
British Archaeologist, 131
bronze accents, 150
Bronze Age, 88, 124, 130, 210
Brookhaven National Laboratory, 21
Brutland Commission, 106
Budh, 44

**C**

Cakchiquel, 161
Calakmul, 159
Calamitous, 4, 155, 157, 159, 161, 163, 165, 167
California, 79, 91, 111, 171, 202-203, 210, 212
Cambrian, 37
Cambridge University, 72, 198, 200, 207, 213
Camille Arambourg, 72
canal levees, 153
*Cannibalism*, 79, 87, 183, 188, 190
Canoes, 10, 183-184, 187-188

Captain *James Cook*, 173
Caracot in Belize, 159
carbon dating, 167
carbon dioxide, 34, 48-49, 102, 193
Carboniferous, 38, 40
Caribbean Sea on the east, 155
carved stones, 179
Catastrophe, 36, 49, 116, 131, 137-138, 193
catastrophic asteroid, 57
cavernous water, 58
Cenote, 58, 166
*Cenotes*, 10, 157
Centipede, 57
Central America, 9, 155-156, 161
centre of spheres, 28
ceremonial value, 179
*cetaceans* fall unconscious, 183
Chandradev, 44
changes in floral and faunal population, 114
Changes in forest composition, 114
Chanhudaro in Nawabshah district of Sindh province, 124
Charles Darwin, 52
Charles Lyell, 53
Charles Watson Boisei, 71
Chechen Itzá, 156
Chiapas, 156, 159, 161
Chiapas in México, 159, 161
Chichén Itzá, 159, 164
Chilopoda, 57
Chimpanzees, 17, 60-61, 63, 65-66, 83, 116-117, 200

China, 38, 58, 75, 84, 91, 126, 193
Chronicle, 11, 154
*Chultans*, 158
Chunchucmil, 159
*Ciguatera* fish poisoning, 173
Cinders, 170
City of Mirador, 161
city of Uruk, 143, 145
Classic Maya Civilisation, 156, 167
classic period, 156, 158-159, 161
clay artefacts, 144
clay tablets, 145, 149, 152, 172
cliff edge at *Val Atare*, 183
*climate change*, 3, 8-9, 14, 19, 29, 80, 82-83, 92, 109-113, 115-117, 119, 137, 141-142, 168, 198, 210
climate forcing or *"forcing mechanism"*, 110
Climatic factors, 94
Coba, 159
codices, 164
collapse of social order, 185
collapse of Sumer Civilisation, 154
*Collapse: How Societies Choose to Fall or Survive, 2005*, 186
Columbus, 91
Comets, 27, 45-46
common ancestor, 50, 60-61, 63, 117
Common Brown Lemur, 59
complex calendar system, 10, 163
concept of carrying capacity, 103-105
Congo, 61

*Conseil Europeen por la Recherche Nucleaire (CERN)*, 26
consequences of global warming, 114
contemporaneous, 164
Copan, 156, 159
Copan in Honduras, 159
Copper, 45, 90, 139, 152
Coquerel's Sifaka Lemur, 59
Core, 14, 31, 42, 153, 166-167
cosmic dust, 23
Costa Rica, 9, 155
court shaped 'I', 162
covering 356 days, 163
Cradle of Humanity, 8, 66, 87
Cradle of Humankind, 69
cranial capacity, 73, 81, 83, 117
Crawford School of Public Policy at The Australian National University, 103
Creation of moon, 45
Cretaceous, 39
Cretaceous-Paleogene extinction or End Cretaceous, 39
critical for food availability, 180
Cro-Magnons, 80, 81
Cro-Magnon 1, 80-81, 83
Crust, 31
cryptogrammic cover, 37
cultural evolution, 155
cuneiform language, 148
*Cyanobacteria*, 36, 48
Cyprus, 88
Czech Engineer and experimental archaeologist Pavel, 182

Czech monk, 54
Czech Republic, 37, 84

**D**

Daimabad, 123
dams, 151, 154
David Harrison, 171
David Nathaniel Spergel, 21
David Otarisdze Lordkipanidze, 76
Decode, 184
Deep Impact, 45-46
deep-rooted crops, 154
deforestation and the erosion of soil, 165
Deforestation in the areas, 100
deletion and insertion, 50
demographic complexity, 185
Denisovans, 86-87
depleting natural resources, 101
deployment of HEMM:, 113
depositional chronology, 166
Desalpur, 123
developed trading links, 126
Devonian, 38, 40
Dholavira 123, 124, 127, 129, 132, 133, 141
Diademed Sifaka, 59
diatomic allotrope, 56
Dilbat, 146
Dinosaur Tyrannosaurus Rex, 57
Dinosaurs, 39, 49, 57-58
Disheartened, 185-186
distinguished class of professional carvers, 181
Dmanisi Man, 86

## 220 | Index of Words

Dmanisi, Republic of Georgia, 75
DNA mutation, 50
*Dolphins*, 10, 48, 173, 183, 188
domes, 170
domesticated animals, 125-126
Donald Johanson, 68
Dos Pilas, 159
Dr. Catrine Jarman of Bristol University, England, 189
Dr. David Frawley, 135
Dr. George Lee, 179
Dr. Jane Goodall, 8, 17-18, 83
Dr. Jason Ur, 11
Dr. Louis Leaky, 75
Dr. Martin Van Kranendon, 47
Dr. Michael Zolensky, 46
Dr. Stephen Mojzsis, 47
Dravidians, 131, 134
*Drishadvati*, 121
drought occurring intermittently, 166
Dubious, 87
duckweed fern, 49
dug-wells, 128, 139
Dugong, 47, 55
Dutch explorer Jacob Roggerveen, 173
dwelling establishments, 156
Dzibilchaltun Yucatan, 159

### E

Eannutum, 145
early classic period, 156
Early Modern Age, 90
Early Stone Age, 88
Earth, 3, 5, 8, 17, 19-23, 25, 27-29, 31-53, 55-59, 61, 63, 65, 67, 69, 71, 73, 75, 77-81, 83, 85, 87, 89, 91-92, 94-97, 99-103, 110-113, 117, 127, 130, 133, 157, 163, 165, 169, 194, 199
Earth's magnetic field, 31, 97
earth-moving machines (HEMM), 101
earthen pitchers, 158
Earthquake, 58, 137
Easter Island, 10-11, 169-174, 178-180, 182-188, 190, 193, 198, 202-205, 208, 211
Easter Island script, 172
Easter Sunday, 173
ecocide' theory, 11, 186, 189
Ecological concept, 104
economic foundation, 154
Edaphic factors, 94-95
Edouard Lartet, 83
Edward Drinker Cope, 83, 198
Edwin Hubble, 7, 20, 24, 207
Egypt, 88
El baut Ceibal aka Seibal, 159
*El Gigante*, 175
El Mirador Piedras Negas, 159
*El Nino*, 111
El Peru, 159
El Salvador, 9, 155-156, 159-160, 162
El Sidron, Spain, 79
Elamite invasion, 146
End Ordovician, 36
End Permian, 38, 40

End Triassic or Triassic- Jurassic extinction event, 39
Endorheic, 166-167
Energy, 13, 20-21, 43, 91, 110-111, 113, 138, 202
England, 74, 189, 191
English, 17, 26, 157, 207
Enigmatic, 169, 173, 177
Enshrouded, 156
environmental aberration, 154
environmental and cultural differences, 156
Environmental carrying capacity, 105
environmental commodities, 102
environmental degradation, 92, 116-117, 138, 165, 168, 190
environmental degradation leading to climate change, 116, 168
environmental stressors, 37
Eocene Epoch, 49, 59
Eridu, 143, 145-146
erosion of soil, 101, 165
Eshnunna Der, 146
Ethiopia, 65-66, 68, 71-72, 74
Eugene Dubois, 75
Euphrates, 9, 88, 143-144, 146, 153-154, 209
Eurasia, 77-79, 82
evaporated naturally, 154
Evolution, 3, 7-8, 18, 20, 22, 24, 26, 28, 30, 32, 34, 36, 38, 40-44, 46, 48-92, 94, 96, 98, 100, 102, 104, 106, 108, 110, 112, 114-118, 122, 124, 126, 128, 130, 132, 134, 136, 138, 140, 142, 144, 146, 148, 150, 152, 154-156, 158, 162, 164, 166, 168, 170, 172, 174, 176, 178, 180, 182, 184, 186, 188, 190, 192, 194, 200, 204, 210-211
*Exosphere*, 29, 96-97
extended drought, 165
external forcing, 111
Extinctions, 36, 38, 40
extrusive igneous rocks, 170

**F**

faces bearing proud but conspicuously enigmatic expressions, 177
fairy moss, 49
famous calendar, 158
farm output, 161
fauna, 40, 48-49, 52-53, 57, 93, 95, 101, 103
feathered serpent, 163
Feldhofer cave of Neander valley, 77
*Felipe Gonzalez de Ahedo*, 173
Fermions, 26
Fertile Crescent, 88, 143
Figurines, 144
*Firozi*, 129
first egg finder *Hopu*, 176
first Mesoamerican Civilisation, 158, 165
Flora, 49, 52, 93, 95, 101, 103, 162
floral and faunal population, 101, 114
Fluctuation in crop yields, 114

Foramen Magnum, 59
formative or preclassic, 157
fossils of spores, 37
France, 74, 79-80, 84, 169
Francois Baron Englert,, 26
Fred Hoyle, 22, 207
French ethnologist Alponse Pinart, 171
French Naval officer *Abel Aubert Dupetit Thouars*, 182, 183
frequency of landslides increases, 101
fresh water, 10
frugivore, 66
fruits like Avocado, 9, 155

### G
Galaxies, 20-25, 27-28, 194
Ganymede, 44
Gauteng province, 69
genital herpes, 72
Genus Pongo, 63
Geological Survey of Western Australia, 47
Geologist Dr. Mark Brenner, 166
Georges Cuvier, 53
Ghaggar-Hakra river, 121, 132
Giant cloud, 24
Giant impact hypothesis, 8, 43
Giganotosaurus, 39
Gigantopithecus blacki, 86
Gilgamesh, 145
Girsu, 146
Gladysvale, 69
global greenhouse gases (GHG), 91

global temperature, 92, 111-112
Global warming, 3, 8, 29, 40, 90, 92, 109-115, 117, 119, 137
God *Makemake*, 176, 181-182
God of death as *Yum Cimil*, 163
God of Flora as *Yumil Kaxob*, 162
God particle, 25
God sun, 162
Godin Tepe, 145
*Gods, Sages and Kings; Vedic Secrets of Ancient Civilisation*, 136
Gola Dhoro in Kutch district Gujarat state, India 123
Gold and Ivory, 139
Gorilini, 63
Goyet, Belgium, 79
Gradual collapse, 164
Gran Dolina, Central Spain, 79
Granite, 31
Great dying, 38
Great Oxygen Event, 35
Great Rift valley, 66
Greenland, Antarctica and the Arctic, 110
Gregor Mendel, 54
Groundbreaking Scientific Experiments, Inventions and Discoveries of the Ancient World, 152
groundwater-table, 99-100, 139, 154, 187
Guatemala, 9, 155-156, 159-161, 209
Guatemala highlands, 156

Gulf of México on the west, 155
Gymnosperms, 57

**H**

Hadar Triangle region, 68
Hadron Collider (LHC), 26
half-bird and half-man, 182
Halogen, 170
Hamelin Pool Marine Nature Reserve, 47
Hamersley Ranges of Pilbara, 48
Hammurabi of Babylon, 147
hampered, 98, 133, 141, 154
*Hanau Epe*, the ones with "long ears", 177
*Hanau Momoko*, the ones with "short ears", 177
Hanga Roa, 169
hanging garden of Babylon, 150
Harappa, 122-124, 127, 132, 136
Harpoons, 10, 188
Heidelberg, 74, 207
Heliocentric, 8, 28
Helium, 22, 34, 96
Heyerdahl and Ferdon, in 1961, 192
Hiatus, 50, 158
*Higgs Boson*, 25-26
high alkali, 170
high ranking personnel, 162
Hindu mythos, 44
Hindu philosophy, 50
History, 9, 18, 29, 35-36, 38, 44, 49-52, 61, 67, 77, 86, 88, 91, 115, 118, 121, 144-146, 164, 166-167, 172, 181, 183, 186, 189-190, 197-199, 202-204, 208, 211-213

History Begins at Sumer, 144
Hiva-Hiva, 170
Hobbit, 76-77, 86
holidays or the 'days off', 153
Hominid, 63, 67-68, 87, 205, 212
Hominidae, 60, 63, 77
Homininae., 63
Hominini, 60, 63, 65, 73, 77, 80
Homo capensis, 87
Homo denisova, 86
*Homo erectus*, 73-76, 82, 86
Homo floresiensis, 73, 76-77, 86
Homo habilis, 73-76, 85-86
Homo heidelbergensis, 73-75, 87, 207
Homo neanderthalensis, 73, 77, 87
Homo rudolfensis, 73-74, 87
*Homo sapiens*, 34, 73, 75, 78, 80-87, 115, 117
Homo tsaichangenesis, 86
Honshu, 67
Horoscope, 151
*Hotu Matu*, 176, 183
Hubble telescope, 8, 23-24
human burial sites, 178
human heads, 172
hunted *Dolphins*, 10
hunter-gatherer groups, 117
Hydrogen, 22-23, 34, 47, 96, 202
hydrosphere, 93-94, 99
hydrothermal vents, 47
Hylobatidae, 63
Hypoxic, 37-38

## I

imperative part, 162
in situ, 167
*Inca Civilisation*, 119, 172, 174
Increase in southwest, 114
Increase in storms, 114
Index fossils, 38
Indian Institute of Technology, Kharagpur, 125
indigenous group, 9, 156-157
indiscriminate cutting and burning of trees and plants, 165
Indonesia, 75-76
Indri Lemur, 59
Indus Valley Civilisation, 3, 119, 121-125, 127-137, 139-141
Industrial activities, 97, 113
influence on ageing, 170
Infrastructure-congestion relations, 106
Insect, 38, 56-57
intellectual, artistic and cultural heights, 161
Inter-tribal battles, 182
Intergovernmental Panel on Climate Change (IPCC), 109
Intergovernmental Science-Policy Platform on Biodiversity and Ecosystem Services (IPBES), 40
internecine tribal conflicts, 175
internecine warfare, 188
*Interstellar nurseries*, 23
intriguing fact, 164
Ion Microprobe, 46

Iraq, 9, 79, 88, 143, 212
Iron, 37, 42, 47-49, 82, 88, 90
irrigation canals, 157
Isin, 146
Isla de Pascua, 171, 185
island Mangareva, 169
island of Flores in Indonesia, 76
Island Pitcairn, 169
Israel, 84, 88
*Itzamna*, 163
Iximche, 159
Ixkun, 159

## J

J.B. Dutroux-Bornier and J. Brander, 192
Jackson Laboratory at Bar Harbour, Maine, Sacramento county, California, 171
Jade, 164
Jared Diamond, 10, 189
Java man, 75, 87
Jean-Baptiste Lamarck, 53
*Jean-François de Galaup, comte de Lapérouse*, 173
Jeff Hoffman, 23
jellyfish, 56
Johannesburg, 69
Johanson, 68, 203
John Burdon Sanderson Haldane, 54
John Marshall, 123
Jordan, 88
Jovian planets, 28

Index of Words | **225**

judicious use of water, 158
Jupiter, 28-29, 34, 44, 46

### K

K-Pg, formerly K-T extinction, 39
K-T or K-pg mass extinction, 58
Kalibangan of Hanumangarh, 123
Kaminaljuyu, 159
Kamoya Kimeu, 72
Kanjetar, in Gir Somnath district, 123
Karijini National Park, 48
*Kavakava*, 182
Kazallu, 146
Kekchi, 157, 161
Kekchi Maya, 161
Kenya, 66-68, 71-73, 76, 87, 205
Khirasara, 123
Kid-nun, 146
Kimberley, 69
kin lines, 152
king of Lagash, 145
King Shulgi of Ur, 146
kings or *"Kuhul Ajaw"*, 163
Kish, 146
Kissura, 146
Klondike region, 91
KNM-ER 1470, 73
KNM-WT 17000, 72
Kon-Tiki museum, 182, 211
Koolasuchus, 39
Kot Diji in Khairpur district, 124
Krishna Paksh, 44
*Kuhul Ajaw*, 10, 162-163
Kuntasi, Rajodi in Rajkot, 123

Kutch district in Gujarat, 132
Kutha, 146
Kuwait, 88, 143

### L

La Ferrassie in Dordogne region of France, 79
la nature, 18, 20
*La Nina*, 111
La Pradelles, France, 79
Labna, 156
Ladakh region of Jammu and Kashmir, 141
Lagash, 145-146
Laguna *Chichancanab*, 166
lake chechancanab, 156
lake Rudolf, 68
*Lake Titicaca*, 172
Lake Turkana, 68, 72-73
Lakhueen-jo-daro in Sukkur district, 124
land of the civilised kings, 143
Land use Change & Forestry, 114
*Lapis Lazuli*, 129, 139, 141
large monolith stone sculptures, 174
largest single block of indigenous people, 159
Larkana, 124
Larsa, 146
laryngeal diverticula, 74
*Last Universal Common Ancestor, "LUCA"*, 50
Late Devonian, 38
late Ordovician, 38
lava flows, 170, 175, 181

LD 350-1, 73-74
Lebanon, 88
Lectotype, 73, 83
legal code, 147
legendary first king of Rapa Nui, 176
Lemuridae, 59
Lemurs, 59
Les Eyzies in France, 80
lime water, 157
Limestone, 58, 69, 71, 79-80, 157, 164
Lithosphere, 93-94, 99
Little foot, 69
local drainage system, 101
Lokeno formation, 67
long count, 163
Long Count Calendar, 163
Long-term and periodic climate change, 112
Loteshwar in Patan district, 123
Lothal, Rangpur in Ahmedabad district, 123
Lucy, 68, 85, 203

**M**

Macaques, 60
Macroevolution, 53-54
Madagascar, 59
Madame Blavatsky, 135
magical spiritual quintessence known as *'Mana'*, 179
*Magnetosphere*, 42, 96-97
Mahabharat, 135
maize, beans, 9, 155, 157, 161
maize, beans, squash, 157, 161

Makapansgat, 69
Malwan, Surat district, 123
man-made cisterns, 158
Manatees, 55
Mandi in Muzaffarnagar district of Uttar Pradesh;, 123
Manganese and iron Oxide, 82
*Manhar, Mansar,* 136
manioc or the cassava, 157
Mantle, 31
Marad, 146
Marco polo, 91
Mari, 146, 150
marine shells, 164
Marquesas island, 171
Mars, 8, 21, 27-29, 34, 41-42
Martin Pickford, 66
Martin Van Kranendonk, 47
Mary Leaky, 71-72, 75
*matato'a*, 182
Mataveri airport, 169
Matsue of Shimane Prefecture, 67
Max Mueller, 134
Maya Civilisation, 10, 119, 155-156, 158, 162-164, 167, 208
Mayan world, 58
Mayapán, 159, 206
Meave Leaky, 73, 206
Mediterranean, 79
Mehrgarh of Kacchi district, 124
Melting of glaciers and ice sheets, 114
Meluha, 130-131, 139
Mercury, 27-28, 34, 41, 44

Mesoamerica, 9, 155-156, 197, 213
Mesolithic, 88, 90
Mesolithic *Ageron Age* or the Middle Stone Age, 88
Mesopotamia, 9, 126, 143, 145, 148, 198, 209-211
Mesosphere, 96
Metallurgists, 152
metasedimentary rocks of Greenland, 35
Metaxytherium, 55
Meteorite, 46-47, 97, 157
México, 9, 57, 155-161, 166, 208, 212
Michael Coe, 206
microevolution, 53-54
Micronesian island, 84
Migration, 82, 84, 135, 137, 146, 171, 210
Mikesell (1989), 107
Mikesell, Blignaut and Hassan (2001), 107
Milky Way, 8, 22, 27, 41, 194
Milne Edward's Lemur, 59
Mineral exploitation, 113
mineral Gypsum, 166
modern gold rushes, 91
modern manufacturing companies, 152
Mohenjo-Daro in Larkana district, 124
Molecular Clock, 37
molecular cloud, 34, 41
Moon, 3, 8, 19-21, 23, 25, 27-29, 32-33, 43-45, 97, 145, 162, 194

moon Goddess was named *Ix Chel*, 162
Mopan, 157
Mortimer Wheeler, 131
mosaic structures, 150
most powerful God, *Kukulkan*, the snake God, 163
Moula Guercy, Los Angele county, California, 79
Mount Kailash, 121
Moʻai (Moai), 10, 172, 174-175, 177-183, 186, 191
Mrs.Plea, 69
multiple rift zones, 170
Mundigak in Kandhar province of Afghanistan, 123
mural paintings, 150

**N**

Naachtun, 159
Nafanjo, 159
Nageshwar, 123
Nakbe, 159
Nalanda University, 140
Namibia, 74
Napoleonic period, 91
NASA, 23-24, 45-46, 207
Naturae, 18
natural calamity vis-à-vis environmental disaster, 165
natural catastrophe, 116, 131, 138
natural pH, 98, 100
Nausharo, 124
navigable rivers, 164
Neanderthal bones, 80

*Neanderthals*, 75, 77-81, 84, 87, 117
near-earth-objects, 44
Neo-Sumerian king, 147
*Neolithic* Age or New Stone Age, 88
*Neolithic* societies, 185
Neos,, 44
Neptune, 28-29, 34
Neribtum, 146
New Guinean, 84
New Stone Age, 88
Nicaragua, 9, 155
Nicholas Copernicus, 8, 29
Nile, 88
*Ninkilim*, 151
Nippur, 146-147
Nisarg, 18-19
Nixtamal, 157
Nomadism, 74
Nome in Alaska, 91
Nordic-esque race, 135
North America, 58
north-western Honduras, 156
Norwegian adventurer Thor Heyerdahl, 182
not even buried, 186
Nutcracker Man, 71

**O**

Obsidian, 164, 177, 190
obsidian flakes called *"mata'a"*, 190
oceanic crust, 31
Odum, 104
Old Stone Age, 88
oldest linguistic record, 148
Olduvai gorge, Tanzania, 71
*Olmec, Zapotec, Toltec, Aztec*, 155
*Olmecs*, 158, 161
Omnipresent, 25
Omnivore, 66
Omo 18, 72
On the Revolution of the Heavenly Spheres, 29
onion shell weathering, 181
Ontario, Canada, 35
open sea, 10, 183, 188
Ordovician, 36-38, 40
organ transplant, 10, 170
origin of life, 33
*Orrorin*, 66-68, 85
*Orrorin* tugenensis, 66, 68
Otto Schoetensack, 207
out of Africa, 79, 82, 86, 208
overexploitation of resources, 138, 184
Overpopulation, 165
Oxford Brooks University, 72
Oxkintok, 159
Oxygen, 11, 34-35, 37-38, 43, 45-49, 56, 95-97, 100, 102, 167, 193, 210
ozone ($O_3$), 97, 102
ozone layer, 31, 56, 193-194
Ozone, $O_3$, 56

**P**

Paaseiland, 10, 169
Pacific Decadal oscillation, 111
Pacific Ocean, 10, 155-156, 169, 187, 189
painted pottery, 144

Palaces, 162-163, 199
Palaeolithic, 81, 88, 90
Palaeontology, 29, 82
Palenque, 156, 158-159
Paleoclimate, 113
Palestine, 88
Palm saplings and nuts, 187
Pandit Vamdev Shastri, 135
Panini, 60
*Paranthropus*, 71-72, 117
*Paranthropus* aethioptcus, 71
*Paranthropus* boisei,, 71
*Paranthropus* robustus, 71
*Paschalococos* disperta, 186
*Pascuan*, 173
Pashupati Nath, 127
Pathani Damb, 124
Pathani Damb in Makran, 124
Pathani Sheri Khan Tarakai in Bannu district, 124
Patrimonial state, 147
Paul Adrien Maurice Dirac, 26, 207
Pazuris-Darjan, 146
PBS of USA,, 79
Peking Man, 75
Percolation, 154
Permian, 38, 40
Peru Govt., 184
Peruvian incursion, 184
Peten, 156, 161
Peter Higgs, 25-26, 208
Petroglyphs, 182-183, 205
*Pezosiren*, 55

Pharmacologically, 170
Phenotypic pattern, 134
Photosynthesis, 35, 48
Pilbara, 48
Pilbara fossils, 48
pioneer inhabitants, 182
pioneer of distant stars, 20
Pipal tree, *ficus religiosa*, 127
Pir Shah Jurio, 124
Pirak of Sibi district, 124
Pithecus, 65
Pitz, 162
Plazas, 162
Pleistocene, 74-76, 78, 82, 170, 199, 205
Pleistocene epoch, 74, 76, 78, 82
*Poike volcanoes*, 170
Pollen analysis, 181, 190
Polynesian rats, 173, 187, 189-190
Polynesian triangle, 10, 169, 183
Ponginae, 60, 63
Poornima, 44
*Pora*, 176
*Portrayal*, 82, 117
postclassical era, 88
Pounded, 168
Precambrian, 36, 48
precision core drilling, 166
Prehistory, 51-52, 71, 88, 90
presence of naturalism, 150
Primates, 59, 62-63, 117, 210
principle of sustainability, 165
Production-social relations, 106

Prof. Robert DiNapoli of Department of Anthropology, in the University of Oregon, Willamette Valley, USA, 179
Prognathism, 81, 83
Prognathic, 66
properly fashioned and ritually prepared, 179
*Prosimians*, 59, 117
Protohistory, 90
*Pu O Hiro* or the trumpet of Hiro, 176
*Pukao*, 175, 178-179
*Puna Pau*, 179
Puuc, 166
Pygmy Chimpanzee, 61
Pyramids, 150, 157, 161-163, 213
pyramids, 150, 157, 161-163, 213
pyroclastic cones, 170

**Q**
Q'markal, 159
Quartz Chalcedony, 139
queen of primates, 59
quetzal feathers, 164
Quiche, 157, 161
Quintana Roo, 166

**R**
rain God as *Chaac*, 162
rains played truant, 141
Rakhigarhi, 122-124, 131-133
Rampant, 173, 184, 191
*Rano Kau*, 170
*Rano Raraku* volcanic crater, 172

Rapa Nui, 4, 6, 10, 119, 169, 171-187, 189-195, 208, 211
Rapa Nui language, 173
Rapa Nui National Park, 169
Rapa Nui, the Easter Island, 169
*Rapamycin*, 10, 170, 210
rationing on the use and supply of water, 158
Ratnagar, 138, 140
Ravi, 136
Raynold Aylmer Fisher, 54
recorders of past environmental history, 166
Red-fronted Brown Lemur, 59
Red-Shanked Douc, 59
redeeming feature, 34, 45, 174
regenerative qualities of soil, 101
region of Zanj, 71
Region Valparaiso, 170
Rehman Dehri in Dera Ismail Khan district of Khyber-Pakhtunkhwa, 124
Relativistic Heavy Ion Collider (RHIC), 21
release of noxious gases, 101
religious ceremonies, 163-164
remains of biotic life, 35
repeated occurrence of Gypsum bands, 167
reproduce exponentially, 187
Republic of Chad, 66
Republic of Chile, 169, 171, 185
Republic of Guatemala, 161
Resource-production relations, 105

Resource-residuals relations, 106
rhythmic chant, 181
rice cultivation, 114
Richard Leaky, 72-73
Rig Veda, 122
Ring-Tailed Lemur, 59
Rise in sea, 110, 114
Rituals, 79, 128, 163, 184
RNA, 56, 210
Robert Bakker, 83
Robert Costanza et al. (1997), 102
Robert E. and Carolyn Krebs, 152
Robert McCormick Adams, 153
Robin M Canup, 43, 44
rock monoliths, 172
Rollers, 181
rolling hills, 170
Ronal Clark, 69
*Rongo-rongo*, 172-174, 184, 202
*rongo-rongo* tablets, 184
rubber and chocolate, 164
rubber ball, 162

**S**

sacrificing captured kings, 162
Sahelanthropus tchadensis, 66-67
Sahelian arid zone, 66
sailors from Tahiti, 171
'Sala y Gomez' submarine ridge, 169
Saline, 166
salt-tolerant crop, 154
*Salviniaceae family*, 49
Samuel Noah Kramer, 144
San Bartolo, 159

Sanauli, Sothi in Baghpat district, 123
sanctity of principally, the head, 177
sand quarry in Rosch, 74
Santiago, 169, 211
*Sar-I Sang*, 129
Saraswati, 9, 115, 121-122, 135-137, 140-141, 143, 153, 194, 199, 203-204, 207-208, 212
Saturn, 28-29, 34, 44
Satyendra Nath Bose, 26, 198, 205, 212
Saudi Arabia, 37
Saul Perlmutter, 27
*Scoria*, 170, 174, 178-179
self-destruction of islanders, 186
Semitic Akkadian, 153
sensitive to hearing, 183
Septentrional area, 156
seven Mo'ai (*Moai*) of similar size, 179
sever starvation, 168
sexigesimal system of depicting time, 152
Sewall Wright, 54
Shanidar Limestone cave in Iraq, 79
Shark Bay of west coast, 47
Shark Bay *Stromatolites*, 48
Sharks, 57, 177, 183
shift to barley, 154
shifting trade routes, 165
Shikarpur, 123
Shimane Medical University, 67
Shinar, 143
shoot 12 islanders, 184

short Ice Age, 112
shortage of canoes, 188
Shortgai in present-day Afghanistan, 122
Shortugai of Takhar, 123
shuffling motion, 182
Shukla Paksh, 44
Sifaka Lemur, 59
Silurian, 36-37, 40
Sindhu-Saraswati Civilisation, 121-122
Sippar, 146
*Sirolimus*, 170
Skjdsvold's excavation, 183
slaughtered mercilessly, 183
Smallpox, 184
Smithsonian, 53, 211
snail shells, 167
soil fertility, 180
*solar maximum*, 113
*solar minimum*, 113
Solar systems, 22
*Sooty Tern*, 176
Sotkakoh, 139
South America and Galapagos Archipelago, 53
south Pacific Ocean, 169
Southeast Asia, 53
southeastern fringe of Turkey, 88
southern Iraq, 9, 143
southern México, 9, 155
Spain, 74, 79-80, 173
Spanish, 157, 171, 173, 209
special territory, 169

spices like vanilla, 9, 155
starved to death, 166
state Ista de Pascua, 170
statues 'walked', 181
Sterkfontein, 69
stiff and bloody opposition, 165
Stone Age, 88
stone tools, 52, 61, 72, 77-78, 80, 90, 144
Stratosphere, 56, 96-98, 102
*Stromatolites*, 35-36, 48
subterranean fauna, 48
sugarcane,, 173
Sumer, 115, 130-131, 143-145, 148-150, 153-154, 199-200, 204, 212
Sumerian Civilisation, 119, 143, 146, 153
Sumerian cuneiform, 145
Sumerian identity, 151
Sun, 5, 8, 10, 21, 27-29, 32-35, 41-43, 45-46, 56, 97, 102, 110-113, 127, 145, 162, 165, 172, 194-195, 200
super plant, 49
supercontinent Gondwana, 36
Superfamily Homonoidei, 60, 63
Surkotada in Kutch district, 123
sustainable development, 3, 93, 104, 106-107
sustainable yield, 188
sustained drought, 165
Sutkagan Dor in Makrana district, 124
Sutlej, *the Shatadru*, 121

Swartkrans, 71
sweet potatoes or yams, 172, 187-188
swivelling and rocking, 181
Syria, 88
system of canals, 151
Szibilchaltun, 159

**T**

T. Rex, 58
Tabasco, 156, 161
Taggish lake, 46
Tahiti Govt., 184
*tangata manu*, 176-177
taro or kalo, 173
Taung, Sterkfontein, 69
Tazumal in Chalchuapa region in El Salvador, 159-160
*Te pito o te henua*, 171
Tell Abu Shaharain, 143
Temple-1, 45
Temples, 150, 157, 162-163
*Teotihuacan*, 161, 164
*Terra Australis*, 175
Thames and Hudson, 161, 206
The Ancient Greek Civilisation, 119
The Aztec Civilisation, 119
the black-headed people, 143
The Chinese Civilisation, 119
the doom, 163
The Environment (Protection) Act, 1986 in India, 93
The *Inca Civilisation*, 119, 174
The King River, 126

the land, 10, 36-37, 39, 56, 88, 100, 135-136, 143, 154-155, 157, 175-176, 188
the land of black-headed people, 143
the length and breadth, 194
The loss of top and sub-soil, 100
The Maya Civilisation, 10, 119, 156, 164, 167
the number system, 158
The Persia Civilisation, 119
the Rapa Nui Palm, 186
The Rapa Nui society, 119, 179
The Red-Shanked Douc, 59
The Rock Art of Easter island', 179
Theodosius Dobzhansky, 54
therianthropic figure, 182
Theropods clade, 39
Thomas Hunt Morgan, 54
Thomas Malthus, 53
Thor Heyerdahl, 172, 174, 182, 184, 188-189, 191
Tibetans, 87
tides in the ocean, 45
Tigris, 9, 88, 143, 146, 154
Tigris and Euphrates in Iraq, 88
Tikal, 156, 159, 162, 203
timber planks, 184
Timothy Bromage, 73
tiny ears, 183
Titan, 44
Tom Gray, 68
Topographic and Geomorphological factors, 95
Topography, 94, 170, 187

## 234 | Index of Words

Toppled, 175, 182
total population of Mayans, 159-160
*Totonac*, 161
Toumai, 66
tour de force, 151
*Trachyte*, 67, 170, 174
Transport and Biofuel burning, 113
*Trappist-1*, 28
*Treaty of Annexation of the island*, 185
tree barks, 164
Triassic period, 38-39
Trioxygen, 56
*Troposphere*, 96
Tso Moriri lake, 141
Tugen Hills region, 66
Tulum, 159
Tuna, 173
Turkish leader Bakhtiyar Khilji, 140
Tutub, 146
*Tuu Ku*, 181
two distinct oxygen isotopes, 167
Tyrannosaurus, 39, 57
Tzeltal, 161
Tzotzil, 161

### U

U.S. archaeologist Charles Love, 182
Ubaid, 144-145, 149
Ukrainian, 135
Ukrainian Helena Petrovna Blavatsky,, 135
Ultraviolet, 56, 102
Umma, 146
UNESCO, 169
unicellular, 8, 35, 50, 56
unicellular organisms, 35, 50
United Nations Framework Convention on Climate Change (UNFCCC), 109
units of 60, 152
University of Colorado at Boulder, Colorado, 46
University of Florida, USA, 166
upper Pliestocene, 80
upright man, 75
Ur, 11, 73-74, 144-148, 153
UR 501, 73-74
Ur-Nammu, 144-145, 147
Uranus, 28-29, 34
Uruk, 143, 145-146
Urukag, 146
Urukagina, 145
Utu-Hegal at Uruk, 145
Uxmal, 156, 159
Uxmal Yucatan, 159

### V

valley of Kangavar in Kermanshah province, 145
Vanilla, 9, 155
Vanmaush, 58
varied climatic conditions, 187
Vasco da Gama, 91
Vedic literature, 8, 28, 135
Venus, 27-28, 34, 41
Veracruz and Tabasco, 161
Verreaux's, 59

## Index of Words | 235

village *Orongo*, 176, 182
vital balance, 98, 165
volcanic craters, 170
*volcanic tuff*, 170
Volcanoes, 10, 170, 179
volume in the business model, 152
Voyaged, 186

### W

Warfare, 161, 164-165, 188-189
Waste management, 114
water fern, 49
waxy coating, 37
weather-related mortality, 114
well-balanced pH, 101
western fringe of Iran, 88
wheeled vehicles, 151
William King, 77
wood charcoal, 139
wooden objects, 179
wooden sledges, 181
working or labour section, 165
World Heritage Site, 47, 169
World Meteorological Organisation (WMO), 110

worship of ancestry, 182
worship practices, 158
worshipping various gods, 162
writing system, the hieroglyph system, 164

### Y

Yale University, 161
Yaxchilan, 156
Yaxha in Guatemala, 159
Yucatan peninsula, 156-157, 161, 166-167
Yucatec, 157, 163
Yucatecs, 161
Yukon in north-western Canada, 91
Yves Coppens, 72

### Z

*Zapotec*, 155, 161
*Ziggurats*, 150
Zinjanthropus boisei, 71
Zircon crystals, 46
Zodiac signs, 151

# About the Author

**Dr. Nikhil Chandra Misra** obtained M. Tech. in App. Geology from University of Saugar, M.P., India; P.G.D.B.M. from Ravi Shankar University, Raipur, C.G.; First Class Mines Manager's Certificate of Competency (R-O) from DGMS, Govt. of India; L.L.B. and Ph.D. from A.P.S. University, Rewa, M.P., India, all except M. Tech. during the professional job. He is a member of several professional institutes. Considered to be an environmentalist with a home in industry, he has authored some acclaimed research papers. The areas of his interest also include feasibility analysis for mineral industry, mineral and mine economics, mine scheduling and mine optimisation, environmental studies and management besides, mineral exploration. Ancient mythological studies, ancient cultures, the classics of Hinduism, Cosmos and related literature have figured as arena of his additional leitmotifs, besides keeping abreast with core subject matters and the associated professional meadow.

**Dr. Nikhil Misra** has worked for M.P. state Govt. as an officer, was assigned to the United Nations Development Project—UNDP Mineral Exploration in M.P. for about three years, working for various minerals and later with JK Cement Limited in various capacities including the Vice President, remained actively involved with other group companies also, having worked on the group's several projects across the country.

He figured among top executives and acclaimed professionals in the states of M.P. and C.G., engaged in private sector. Under his leadership JK bagged the prestigious "Quality Excellence award for Best Exploration Project" in the country, instituted by World Quality Congress for a project in Panna, M.P., India for the year 2014. He is also the recipient of a "Life Time Achievement Award" in the year 2017 for his contribution in the field of Mining Geology and Environment. He has been in profession for about 39 years, and has now undertaken writing.

www.ingramcontent.com/pod-product-compliance
Lightning Source LLC
Chambersburg PA
CBHW030917180526
45163CB00002B/368